Falling Into Rishmore

Book 2 of The RishMore Saga

K.S. Riggin

Table of Contents

Prelude

I, Prana, was once a Terran, then more or less kidnapped and forced to become an alien's wife. I'd grown to love my captor husband. Dar had given me two sons, then as the years went by, he'd raised my position from lowest wife all the way to High Two.

He'd made me second in command, which meant I was now the ruler of all.

Dar's death was devastating, but I had no time to mourn him. Right now, I must walk in Dar's footsteps, do what must be done. I must establish my position and protect my sons. Grief could not take precedence.

Chapter One

Prana

Captain Jertot, having gained entrance to Dar's — I mean — my ship's quarters, bowed low. He looked pale, his face crinkled with stress, but he'd come to me, proclaiming me a worthy individual, if not yet giving me the formal allegiance to my position.

Good to know that Jertot might support my succession. Before his death, Dar had bullied the other RishMorian officials into acknowledging my position. Dar had assured me that the nobles would accept me in a few years, but that wasn't something I was offered at the moment.

"I need to know," Jertot said, his eyes downcast in nervousness. "Do I turn the ship around, High One? Do you seek revenge for the death of your husband, our former leader?"

Jertot didn't need to remind me that my husband had been High One. But formality was the tune of the RishMorian. Protocol was everything.

And he'd called me High One. I now knew that I had Captain Jertot's loyalty. It was the first deep breath I'd been able to take since my husband's death was announced.

I nodded. "Yes, of course. I must walk in the footsteps of Dar, the High One.

My tears had probably made my face an angry mask of grief, but such trivialities could no longer play a role in my actions. *Tears were the sign of weakness.* Dar had told me often, even when tears had often swayed the harshness of his punishments. But that was history.

The past remains a mere history lesson, the reflection of triumphs and failures, while the future is an unwritten challenge.

Yes, I hear you, Dar. The imprints in my mind still ring from your lectures, your bombardment of teachings to bolster my abilities.

Now he is gone, but the echoes of his wisdom remain.

My husband had of late threatened to send me into a council meeting alone. A test that would have sunk me, or at least I'd feared that it would have. But now, the threat had been replaced with reality. I would no longer have any choice.

"Yes, we shall return to Crezand," I repeated, my voice, firm and level as Dar's would have been. "Crezand will pay for the High One's death."

My eyes were stinging from recent tears. I felt strange and discombobulated. Zoff, the ship's medic, had warned me that my body was going into shock. He'd wanted to medicate me by spraying me with the pleasant sleep-producing vanilla drug. No time for that, not with a ship at emergency status. I had sent the medic away.

Captain Jertot's eyes flared with appreciation for my words, although he gave no opinion, another recognition of my status — one not to be questioned.

But Dar had given me other secrets with his constant instruction. I could read the captain's thoughts on the matter through gestures, body position, and facial tics. And then, there were my ESPer senses. Long delayed by my upbringing, its suppression in the Center, a prison-like Terran school where Earth big shots had kept me castigated and ashamed of my talents, I could feel my gifts returning.

I knew as surely as if Dar were beside me that the captain approved of my words, my delivery, and the way I was holding in my grief.

I rose from the bed and walked to the ship's monitor. "Damage report. Casualties," I ordered.

As the monitor's voice tonelessly read off the list of names of crew who'd been killed, it finally reached the end one: High One, Dar…"

I sighed and swatted angrily at my tears. "What was the damage to the ship?" I asked, breaking off the recitation of Dar's titles.

The words that followed were less painful: destruction in the dining hall, the main kitchen, the warlord's hall, the baker's ovens, the…

"See that it is taken care of," I ordered. "Assign directives and deliver them to me at once. Put Engineer Stofle in charge."

Once more Jertot nodded his head, approving my actions. He was still standing, stiff and uneasy. Frick, my Sentinel, stood at the door, ever watchful. Other guards barred the way outside the entrance to what was formerly Dar's huge suite of rooms. I guess I'd have to start calling it my…"

Somethings couldn't be thought about right now, like loneliness and grief. Like a heap of responsibility on shoulders that hadn't even seen adulthood yet on my home planet.

But Dar had constantly told me: *When you don't feel qualified yet, my wife, you must simply fake it.*

I'd be doing a lot of that — pretending I was sure of myself, play-acting that I knew how to be the RishMore High One.

But that meant starting with Jertot. I pushed my nose in the air, gave him a level look — an assured look. "When do we reach Crezand?" I asked. "What preparations do we need to make for war?"

"Three days. With your approval for our turnaround, the generals have already initiated war formulation. They will be drawing up plans to submit to you as we speak."

"Good," I said, although inside I was quivering. War, generals, proposals. Could anything be more terrifying? I could see the haughty males in my mind's eyes, bending over the table, their eyes fastened on figures, statistics, and the best impact for the doom and gloom of a million citizens. And in between those calculations, would there be whispered remarks and doubts about me?

I turned to face my Sentinel. "Frick, have someone send for Shnize to report to Dar's — to my chambers. His presence is required."

Captain Jertot awaited further orders. His blue ears dangled in submission. His eyes were red and grief torn. I should soothe him, but I could barely keep myself together. I didn't have even a spoonful of magnanimity for his anguish.

I had done all I could think of. "I am glad you were not in the dining hall," I told Captain Jertot, digging deep for compassion and patience. "Did you lose many of your officers?"

He sighed. "We will survive this attack, High One. There have been others. Dar's father was killed similarly. However, if you will forgive me for asking, but I need to know — are you prepared for what will come? Battle is not pleasant, and you are, perhaps, inexperienced in such gruesomeness."

I took a deep breath. Why had this all happened? Why hadn't Dar protected himself better? Why had he…? Tears poured down my cheeks, but I pretended not to be aware. I quickly wiped them away and continued with what I knew I must do.

"You will prepare me for what I must face, Jertot. You will inform me each step of the way the role I must take. Assign good men to help me. I will not let my husband's memory down. I will uphold the honor of the RishMore."

Jertot smiled again and gave me an approving, formal bow. My words about the 'honor of the RishMore' had curved his lips, but, they had also brought forth his nod.

“Gladly, High One," he said gently. "I will give what guidance I am allowed. May I ask, though, why you have sent for the Franian, Shnize? Do you have concerns about the welfare of your sons?”

I stood up to pace. How much could I tell Captain Jertot? How much information did I dare give him? But maybe, the better question should be, why *had* I sent for Shnize? I wasn't clear about that. I just knew I needed him.

"It is a personal matter, Jertot, but I require Shnize by my side from now on. I trust him. He will assign others to my sons. They are safe now that the Crezand traitor is dead.

“Dar’s death was quick, Jertot, was it not?” I asked, leaning my head against the wall where my husband had long ago placed a picture for me, a picture of a tree.

“He felt nothing, High One. It was a sizeable bomb, and he was next to the traitor. There would have been no pain.”

I breathed in slowly. Then out again. Breathing had become difficult since the hours of Dar’s death. I’d never thought about it before. It used to be automatic, but now I found that I had to think about it, or I forgot.

A knock sounded at the door. I didn’t question who it was. I knew it was Shnize. “Enter,” I called out. “Come forward.”

Once Shnize had been the very image of my enemy. He’d been my guard, my keeper, and my jailer. Then, time had changed me and, maybe, it had changed him. Shnize was a Franian, a cell-dweller, incapable of feeling more than duty and allegiance, yet he was my friend. And above everything else, I needed a friend, for without Dar, I had no one.

“Leave us now, Jertot. But keep me informed. I must be trained. I must be ready,” I commanded.

Jertot shot a sharp look at the huge guard, saluted me, and then retreated. Then, I turned to look up at the one I had sent for.

"I am removing you from the nursery," I told him. "You will assign the most qualified, the most dependable of guards for my sons and for the Franian boys. Henceforth, you will remain at my side."

"I do not understand," Shnize said stiffly. "You are not content with my work in the nursery, High One?"

Franians do not question authority. They only obey. I darted a look of surprise at Shnize's face. Was that a twinkle I saw in his eyes?

I shook my head. I must be imagining things. I knew that Shnize had always resisted my offers of friendship. He would certainly never dare question me.

"Sit down," I ordered. "Of course, I am not dissatisfied. You do everything perfectly, but now, you must be the memory of Dar. You must teach me everything he did not have time to instruct me. You must impart everything he knew, everything you observed, all that you learned about his history, his wisdom, his beliefs. Do you understand?"

"I am not ready for this position, Shnize," I attempted to explain. "I did not want it, but I must be the High One. I must make Dar proud of me."

"It is a considerable load for one so young," Shnize dared to say.

I twisted about to stare at him. "You question the High One's judgment? It is he who made me so."

"I question nothing, High One. I am Franian."

He was playing with me. He had done so often. I knew the game, but yet…

"Please don't say that, Shnize. I can't take it now. Please," I said as I collapsed onto my bed with tears that would wait no longer.

Shnize

With her eyes brimming with tears, she tells me to stay at her side. My right cheekbone twitches in protest over the tension of the moment. I cannot be Prana's Dar. I could never order her, slay her with a word, or command her respect with clever phrases. I am only a Franian, bound to her through allegiance and duty. I cannot be more.

Her tears continue. My arms ache for her. Does she not know that I would grovel at her feet if only she asks? I would give my life for her, but, of course, she knows that. That is my duty. Yet, I would give her even more than that. I would offer up my honor.

How can it be that I suffer from these emotions? Franians do not feel. Even for our Queen on the mating night, a Franian does not shelter the emotion of love. Lust, perhaps; power, but never an emotion of depth.

Yet, I love Prana, and I feel this since her arrival — from that first moment.

I move toward her bureau and forage there for a kerchief. Before, the High One always provides them for his wife. She has frequent need since he makes her weep so often. I carry the kerchief to her and toss it down.

"Thank you, Shnize. Dar shouldn't have died. I should have been there. Maybe Mada wouldn't have bombed him if I'd been there. Maybe…"

"You waste time in what ifs, unless self-abuse serves your need to grieve," I tell her with less gentleness than she deserves.

"Self-abuse?" she repeats, looking up at me with surprise. "But it was my fault. Don't you see? You wrote me that letter, Shnize. I should have listened. You were right. But I ignored it. I stood up to him. I disobeyed him. That's why it happened."

I look up at the wall, fighting for control. I cannot let her see what her tears do to my soul. "His death happens because it is his time, High One. One cannot avoid one's destiny," I say as gently as I can.

"How do you know?" she replies with a sharp burst of rage. "Is that Franian philosophy? Is that what they taught you before they shipped you off to be Dar's slave?"

She is angry with herself and needs someone to attack. I accept the position, yet I know that such a refrain will not help her go forward.

"I am not a slave, High One," I tell her. "I am merely bonded."

She looks at me with eyes like the Franian Sea. Each of her eyelashes holds teardrops. They sparkle in the artificial light of the room. She takes my breath away with her beauty so much that I am grateful it is not necessary to speak at that instant. My words might not be fitting.

"I'm sorry, Shnize," she apologizes, sighing heavily. "I'm half-mad today. I cannot think. I can't speak without bitterness swelling my tongue full of lies. Forgive me, please."

That a High One should apologize to a lowly Franian is beyond madness. I shoot a glance at the door where Frick is watching. He glares at me with ardent hostility.

"Blow your nose, High One. You will feel better soon," I instruct her, attempting to shut out the guard's look, but Frick is right. I am not treating the High One as her position deserves, yet I cannot turn away from Prana, nor can I retreat from the path that we previously forge.

For once, she obeys me, blowing indelicately into the cloth. Then she laughs. “You’re right, Shnize. You’re always right. I do feel better. All I needed was a good cry and a hearty nose blow.”

She falls back onto the bed, stretching her arms over her head. She still clothes herself in Dar’s colors. I wonder how long she will wear them. As High One, no one restricts her. She is now free of his choice.

With her full-body stretch, her legs uncover. The sight of them disturbs me. I look away, attempting to block out the intensity of the feelings inside me.

As if she suspects, she tugs at her dress, pulling it back into modesty. “Sit down,” she orders, patting the bed. “It is good to have a friend at my side.”

I sigh. “The guards would not like that,” I tell her, uneasy with her command.

“Close the door, Frick,” she orders abruptly, shooting glares like orders.

Her primary guard, her former guardian with Dar, bows and without a word of reproach, exits and waves the door shut. He could not do that if Dar still knows life. Then, she could not be alone with me.

As if that occurs to her, a fresh wave of tears follows her command.

“What would you have me do for you?” I ask, trembling a bit because I want her to command me to hold her.

“There’s nothing you can do. Just be my friend. Please. Stay by my side. I trust you, Shnize. There's nobody else now I can trust. He’s gone.”

A long time ago, her tears pour wet on my chest. I hold her then to relieve her fear. Dar, her husband, almost kills me for doing so, but then in the end, he only laughs, knowing I am no threat. Franians are neuters, you see. Everyone knows that. Yet, my organ does not, and my heart, beating with the song of love, does not recall the fact either.

"I cannot stay here," I tell her. "It is not proper. You should not order the door to close with me inside."

"I won't let you leave, and I'm not opening the door again," she says, with her wet eyes flaring with anger. "I want my privacy, and I need you." She reaches out and touches my arm.

My hands ball into fists. I grind them into the material of my uniform. I shut my eyes and attempt to think of other things, but her body lies on the bed. That is all I can focus on.

"Prana," I say, before I realize her name is even on my lips. "Forgive me, High One," I beg at once. "I do not mean to call you that."

She looks up at me, and a half smile trembles on her pale, pink lips. "Thank you," she says. "I want you to call me that. I want to hear my name. No one ever says it now. No RishMorian ever called me that, except Dar, and now he's gone."

Once more her torrent of tears arrives. I stand up. It is all I can do not to seize her in my arms and feel her tears bathing my chest with her fragrance again.

Prana

I wiped my tears and sat up. My dress was twisted about me, so I pulled it back down and straightened it. Shnize's eyes almost jumped out of their sockets. Was he that straight-laced? I'd only showed the tiniest bit of thigh. I displayed more than that in my old Terran shorts. I wondered what Shnize would think if he saw me in a bikini.

I sighed. What was the point of thinking like that? Dar had made it so I could never go back to Earth. I was bound forever now to RishMore sensibilities.

"Shnize? Would you go back to your home if you could?" I asked him.

He stared into my eyes, almost as Dar used to do. Was Shnize trying to read my thoughts?

"My home is with you," he responded. "You hold my allegiance, High One."

Shnize's voice seemed distant. I was used to the sometimes odd way that Franians avoided past tense. Dar explained it to me once. *To speak of what is gone, what is past, is to invite Death to walk with you. That is their belief and the custom of their speech.*

I accepted the fact that Shnize was an alien, a non RishMorian, and his cell background prevented his sharing emotions, but couldn't he be less stiff, less formal? Angrily, I bounced up.

"You're not answering me. If you weren't bound to me, would you go back to your home?"

Sometimes Shnize had a slight twitch in his jaw. It jerked then, fluttering like a butterfly. Did I make him nervous?

I walked closer. "The door is shut, Shnize. You don't have to fear that we're being spied on. Tell me the truth. What would you do if you had total freedom?"

"High One, I ask permission to return to the nursery. The boys will be waking up from their naps now. There is no time for a replacement, so I feel that it is my duty to return to them. Will you grant me that?"

My boys. I had forgotten. Of course, I must release Shnize. The children would be worried. I wondered if they'd already heard about their father. Would they care? Had he impacted their world enough that his death would be a loss?

"I will go with you," I said.

Shnize bowed in response and turned to face the door. Together we went forward, the side of my shoulder almost touching his arm.

Shnize waved the door open. Frick stood there, of course, running his customary body sweep over me, his eyes checking that all was well. I wished that I could tell him to take the day off and leave me alone, but, I couldn't. My guards had been doubled since Dar's murder. Frick said that was necessary in order to reassure the crew that I was safe.

As I stepped through the door, fifteen Franian guards surrounded me, all of them so tall, they completely shut out my view. Their bodies smelled of musk and maleness. I didn't object to that, but it would have been nice to occasionally breathe fresh air.

"Have the guards stand back away from me," I ordered Frick, and without questioning my order, it was done. No argument, no checking with my husband, who would have refused, as he'd always done.

I sighed and sniffled.

The halls were deserted. People had been warned to stay inside their rooms. The heavy, military stride of the Franians pounded the beat of our journey. I scuffed a shoe, trying to keep up, walking with the extra skip their long strides forced me into.

"Slow the men down," Shnize ordered Frick.

The two Franians glared daggers and swords at each other, but Shnize held the gaze longer, his eyes almost murderous with rage. It was Frick who looked away, and the cavalcade that included me in a semi-squashed center, continued at a slower rate.

I put my hand on Shnize's arm to get his attention. He shuddered, but he didn't pull away. "Thank you," I told him warmly.

"I prefer that the High One not touch me," he growled so softly that I'm certain the others didn't hear.

I dropped my hand, but I still teased him. "I thought I had the power to do whatever I wanted now," I said, echoing words he'd once told me concerning Dar's position.

Again Shnize growled, and then he looked down at me. "You are free to torture me, High One. You can order my death, too, although you then need to pay my cell queen for the taking of my life."

This time it was I who shivered. It was easy to forget that the Franians were not like us. I was silent the rest of our walk, wondering why he'd spoken to me so harshly.

As we arrived at the nursery, I discovered that the boys were indeed already awake.

"Mommy," cried out Frawn Thorwain Shadoc Stolk, my older one. He ran to me and clung to my legs.

"Mum," yelled out Strandor Urbaf Shadoc Thraft with a piercing scream of a greeting. He sat up and held out his arms.

I smiled. Then I sat down on the soft carpet and let my sweet boys climb into my lap.

"The High One does not allow that," Frick reprimanded me before he realized that I could now do whatever I wanted. His face turned pale, and he bowed clear to the ground. "Forgive me, High One. I shall resign my post immediately." He spoke with a trembling voice, his body cringing from his error.

A Franian will stay in such a stooped carriage as long as it takes for their master to order them back into guard position. "Frick, get up," I said. "I do not accept your resignation. I desire your protection. Therefore, I shall ignore your audacity — this time."

Frick gave me another deep bow and then stepped back. I ignored him and smiled down at my babies. I gave them both noisy, smoochy kisses. Strandor, as usual when I did something irregular like that, went into giggles. Frawn looked at me with the stern, questioning look of his father.

“Mother, is that allowed?” he asked.

I laughed. Then I tickled his sternness away.

“Frick, Shnize, did you see that expression?" I asked, suddenly, laughing. "Frawn is so like his father. Did you see his eyes and that look he had? The lips, too. He smiles just like Dar did. Don’t you think?”

Frick’s face was expressionless. Nor did he answer me. I might as well have been speaking to the wall.

“You are correct, High One," Shnize responded. "Your son is very much like his father. Will you thus cast off the RishMore training that he needs?”

“Shnize,” I protested, and then I stopped. He was right. I should consider all my interactions with the boys. Frawn would be the ruler of the RishMore one day. Strandor would govern my planet, Earth.

Yet, not this day. My beloved husband had just died. I needed the hugs of my boys. I needed to kiss them, laugh a moment, and forget the pain inside me.

Just as I was thinking that, Frick broke in with a nasty and possibly violent reproof. “You will not address the High One in that manner,” he growled at Shnize, touching his weapon with a readiness to fight.

“Halt. No weapons,” I ordered. "And do not question Shnize's words to me. Not ever. He is…"

I stopped, unable to say what I wanted to say, about how he was my friend.

“Frick, you will remain my chief guard, but Shnize will be my… What should I call him? He couldn’t be my chief guard any longer. Dar had disallowed that, and I would not overrule my husband’s wishes. Besides, I wanted Shnize at my side, available to talk with me, not on duty in the stiff posture of his position.

"He's my — my — Prime Minister," I said.

"I do not understand," both males responded.

"I will discuss it later," I told them. When I've had time to think it over, I said to myself.

What responsibilities could I give Shnize, and what should I tell the cabinet about his position? How could I handle giving him ANY position when he was merely a Franian? Then, I remembered the important thing. I was the High One. I could do whatever I wanted.

Frick went back to his position at the door, and I was left with the questioning eyes of Shnize and the curious looks of my sons.

I shut the door on that worry and played with them. Frick and his team kept guard at the door. Shnize stood watching me.

"May I have permission, High One, to discuss the nursery situation with another of the Franians?" he asked suddenly. I nodded, and he left. Of course, I couldn't deny him that. I needed him to find a replacement. But it was odd how much safer I felt when Shnize was near.

When he returned a few minutes later, his face was as inexpressive as always, but I saw the twinkle in his eyes. He was laughing at me because I'd just caught the soft, fluffy Chism that Strandor had thrown. Sure, I landed on my stomach, but I'd caught the silly thing, and my son was pleased as a non-molting doltor.

"Does the High One wish assistance getting up?" Shnize asked, stretching out his rough-textured, huge warrior's hand.

"If she needs assistance, I will provide it," interjected Frick, as he strode forward with a face so dark with anger, it was like looking at the underside of a rain cloud.

"No, thank you, Frick. You see, that is one of the duties of a Prime Minister," I told him. "Shnize must always pick me up when I fall down."

I could tell that my answer amused Shnize. He was having difficulty not giving in to his very unFranian sense of humor.

"I shall be very busy; will I not?" he asked me, his lips quivering with an almost imperceptible spasm of laughter.

His words sent my older son into hysterics. Frawn, in a sitting position, absorbing everything as always, suddenly fell backward onto the carpet, giggling loudly.

Shnize looked at him, raised an eyebrow, grabbed my hand and pulled me up.

Frick, once more standing at the door, looked on, his grave face, a mask of blankness.

Chapter Two

Shnize

I do not understand what she intends for me. She confuses me with her label of Prime Minister. I think it is an artificial title. Prime means first or most important. Yet, that cannot be her meaning. It also refers to numbers that are indivisible. Is that what she means? Or perhaps it is a defensive term. Am I her main sounding board? That seems closer to what she desires — that I listen and occasionally speak.

I puzzle over it as I watch her with the boys. She invites her Franian children to join her in play, as well. The older ones stand back. They know already that it is not their place. They accept their position. Will Prana undermine that? Will she break down the little ones' training?

I shall speak to the new nursery master, Samuin. He will make sure that her unorthodox behaviors does not upset the system. The young Franians are to be her guards, not playmates for her boys and herself. She will understand that eventually, but she is still young. She has no time, yet, to season in the path of the RishMore.

Frick puzzles over the odd behaviors of the High One. I can explain her lapses to the guard. I believe it is the behavior Prana learns on her home planet before she becomes Dar's bride.

Before, I spend time studying such things to enhance my understanding of her, but I will not suggest Frick's exploration into such areas. He must draw his own conclusions or, probably, not even think about it. That is the more suitable Franian conduct. It is I who steps out of the norm.

And why do I? Dar says that time clarifies better than words. He does not speak then of his wife, but it aptly fits her. To be near her, to watch over her, to feel the great heaviness of love — that is the clarity that time brings. No greater understanding is there. Love sees into the soul.

Look at me, a Franian guard, and I dabble with such thoughts. I, too, need to listen to the training manuals. I must remind myself constantly of who and what I am.

She kisses Todd. His eyes study her. That confuses him. He is secure in his programming. He knows he is merely Franian and not one of her sons. Why does she kiss a guard-foster? Does she not see the confusion? No, she cannot see. She laughs and plays. She does not see the damage.

Todd watches her. I see him ponder it. I take him to the side. "The High One lacks her husband. Bereavement and confusion attack her. Do not allow her kiss to unsettle your purpose."

I speak into the boy's ear, and he listens attentively, quietly, and calmly. Frawn would not. Frawn would rebel against such words. He would throw his head back and glare at me. Already he feels the royal blood burning his soul. He is conscious of it. Todd must be equally conscious of his lowly position. He must remember he is only a Franian.

I attempt to explain that to Todd, but Prana calls him over to her. He goes reluctantly, his sideways glance at me, a mixture of eagerness and dread. Ah, Prana, I wish I could interfere. What you are doing is not a wise thing.

Prana

Shnize frowned at me. So did Frick, but not as pronouncedly. The latter was far more discreet, but I felt the disapproval of both of them.

Yet, I have waited so long to undo the harm that Dar did to these poor children. He never allowed me to hold my Franian guard-fosters. But nothing can stop me now. I shall foster them with love.

I do wonder if Shnize will lecture me later about it, though. Would he dare? I told him that he must be open with me, so if he scolds me, that would be okay. I would prefer that, in fact, to his present coldness.

My older son, Frawn, is growing so fast, but he was also stern lipped with anger because I encouraged the Franian boys. He will get over it. Little Strandor, however, was delighted. He laughed and played with the Franians as eagerly as I did. His prejudice has not yet been cultivated. I shall make sure it never is.

I played for several hours. Then, because Shnize pulled my attention to it, I realized that I was tired. I had not eaten yet either. He recommended that I let the boys take their lessons. So, we left the nursery.

I was willing to follow Shnize's advice. At least I would try to eat, even though I felt little hunger. We had not gone more than three halls down the corridor when Captain Jertot approached. Bowing deeply, he said, "High One, I ask that you grant a moment of your time to inspect the reports."

Of course, I must do so. I followed him to the command room, where he'd spread them out as if they were treasure maps. Then he began to pour over them with analysis and ship talk that was so far beyond my comprehension, I simply raised my hand to stop him.

"Jertot," I began. "I am lost in your sea of words. I don't understand many of them. Please, can't you talk slower? The language is still difficult for me at times."

"Forgive me, High One," he said, dropping down to his knees. "You may have me beaten if you like. My failure to speak to your understanding was inexcusable."

The captain was a proud RishMorian, respected by all, yet at the moment he cowered before me, like a bluish puddle at my feet.

"Don't do that," I told him, reaching over to pull him up.

"No! You mustn't," Shnize snapped at me, and then he, too, dropped.

"Both of you get up," I said, sighing heavily. "Shnize, I've asked you to advise me. If I do something wrong, I appreciate your corrections. Please. Both of you. I am still the same person I was. I need your help."

Shnize rose, but his eyes were still lowered. "I speak out of turn, before, High One. I will return to the nursery if it pleases you."

"No, and stop that," I said again, getting angry at the silliness of the situation. "I've told you that I value your words, Shnize. Jertot, I order you up. This is ridiculous. I will not have the two of you dropping down on your knees whenever I do something wrong."

Jertot rose, his eyes not on me but on my guard. Frick had drawn his weapon.

"Put that away," I yelled at Frick. "You are never to draw a weapon on either of these two, RishMorian and Franian. Do you understand?"

Growling under his breath, the guard returned the weapon to its holder. "I anger the High One?" he asked.

What a mess. I sank into one of the chairs, put my face in my hands, and tried to think calm, cooling thoughts.

All the males remained standing, waiting for my next command.

"I can't do this," I moaned. "I just can't. Dar was wrong. I don't have it in me to govern."

"Shut the door," Shnize ordered. Without a word of argument, Frick complied.

"You are the High One," Shnize told me, his voice, softer than it had been all day. "There can be no abdication. You cannot go back to the before. You must take the command and not worry that you error. Demand. Order. Rule. Do not question yourself. It is unseemly."

Jertot was staring at Shnize, his mouth open, his eyes two-sizes bigger than normal, and his ears flapping with worry and tinges of anger.

"I have made Shnize my Prime Minister," I told Jertot quickly. "He is required to speak freely. I have commanded it."

"I see," Jertot said with the slightest bit of a flicker of amusement in his eyes. I saw that I would have no problems with Jertot. He would not question my friendship with Shnize.

Then I turned and smiled at my former guardian. "You are right, Shnize. Thank you.

"Could you please explain all these details again, Jertot?" I asked, standing up and walking toward the two males.

Slowly, very slowly, the captain began. Shnize broke in several times with explanations that helped as well. Tired as I was, it ended up a fruitful afternoon.

When at last, I let out a rather large yawn, Shnize picked up the diagrams I was staring at, curled them into their cylindrical keeper, and handed them back to Jertot.

"The High One must rest now," he said. "And she must eat."

"You will not…" Frick began, but I glared him into silence.

Luckily, my guard didn't fall down to the ground with the rebuke. Instead, he shifted to an even more rigid position, indicating his extreme anger. I sighed heavily, nodded my head, yawned again — this time, not even bothering to try and cover it up, and followed Frick as he, military style, made an abrupt turn and headed out the door. Jertot remained in the room, gathering up the rest of the plans, but

Shnize, moved immediately behind me, so close, he was almost stepping on my heels.

I found myself suddenly so tired, I didn't have the energy to argue about such closeness, even if I'd wanted to. The other guards moved to shut me inside their bodies of protection, which brought on my usual tendencies of claustrophobia. I yawned a third time, my tired eyes watering so badly, I stumbled.

Shnize reached out and grabbed my arm. "I will carry you, if you wish," he said.

I shook my head. "I must not appear weak," I said. "Dar used to tell me, that even the floor has eyes."

"That is true, High One. It is better if you do not need me to carry you," Shnize said gruffly, dropping his hold on my arm.

Nobody said anything more. I was certainly too tired to keep a conversation going. I studied the floor and watched my feet move one step after another.

When we reached my room, there was no hesitation on Shnize's part about entering. Nor did he ask permission to close the door behind us.

"Sit down," he said. "I will program a meal. You must not sleep until you eat."

I nodded gratefully, but I leaned against the wall that butted my bed, then closed my eyes. "I miss him so much," I said.

"Of course, you do. Your husband, the father of your sons, he is, before in life, your all. It is fitting that his absence makes you sad."

"He always made me do things I didn't want to do. He never let me just be myself. He wanted me to rise into power. Why, Shnize? Why?" I was sniffling again, warn-out, grieving, and desolate from the bombing and the death of my beloved husband.

Shnize chose a meal for me. Then he brought over some soup, hit the button for the table to lower, and set the bowl down in front of me. Limply, I sat up, not really caring if I ate or not, although I had to admit the soup smelled amazingly good.

"You are the only one he gives his trust to," Shnize told me. "Of all his wives, only you have the intelligence, the courage, and the brain touch."

"Brain touch?" That was a new one to me. I laughed, but it petered out from fatigue. Besides, it didn't feel right to laugh, not with Dar gone. How could I ever feel light enough, know contentment, be jovial, laugh…without him? And how could I ever go on pretending that I could be the High One? Attempting to act like I knew what I was doing?

"Eat," Shnize commanded, seeing my sudden sadness. "You will feel better then."

"You have not eaten, either," I said. "I'm not going to eat unless you do."

Shnize stared down at me. His face tightened with disapproval. His posture, rim rod straight, suggested that I'd said something traitorous. "I cannot sit and eat with you," he told me. "That is impossible. I am not your equal, High One."

"Then I will eat nothing," I told him firmly, closing my eyes and pushing the table away.

Shnize stomped off. I thought he'd decided to leave. The silence was good, though. I welcomed it. I was still sitting up, but I leaned back against the wall. I was half-asleep when Shnize slammed another bowl down on the table.

"All right," he said. "I will sit and eat with you, High One, but only this time. It is not proper for a Franian to eat with a ruler."

I snickered. "Ruler! Don't be silly. I'm just a Terran girl who somehow got mixed up with the RishMore. I'm no more a ruler of worlds than you are."

"You will not say that again," Shnize said, pointing his utensil at me. "As Dar says to us often, the floor has ears."

"Eyes," I said, chuckling in spite of myself. "The floor has eyes."

Shnize pulled up a chair and sat down. Then, he handed me the spoon. "We bargain, remember?"

I took the spoon, dipped it into the bowl, and began to eat.

I think Shnize watched me more than he filled his mouth, for his soup bowl was only half-empty when I was finished, but he stood up, then, and went to get the rest of my meal.

"I won't eat anything else, either, unless you do," I warned him, smiling a bit, because I really did enjoy teasing him.

Shnize humped loudly and then came back, carrying a dish of meat and vegetables, which he placed in front of me. He picked up my utensil again and shoved it into my hand. "Mine is cooking," he said. "Eat this before it gets cold."

Strangely, the soup had encouraged my appetite. I took a bite of the zepers and then finished them off. By then, Shnize was bringing over his meal. He set it down on the table, looked at my plate, and smiled. "It is in my memory that you like them," he said, looking pleased that I'd gobbled them up like a starving dog.

"What will we do to Crezand?" I asked, breaking into new ground.

Shnize readjusted himself down on the small, blue visitor's chair. His huge body seemed ridiculous on the tiny, rounded platform. I started to suggest that he sit beside me on the bed, but I held my tongue. What would he think if I suggested that? Would he find me brazen? Would he, frankly, just refuse?

"We must destroy their major city, High One. It is what Dar would do. If we do not retaliate, others will strike back at us. Such an unraveling of protocol is dangerous. It cannot be otherwise."

"I don't want to kill anyone, Shnize. I never have."

"The freezing will not kill them, High One. But do you wish to endanger your sons? If you do nothing, you may destroy a thousand worlds for your simplistic beliefs. Is that good?"

"Is there no way we can reason with the Crezands? Maybe we can compromise…"

Shnize's face had turned warrior. I think, if he'd dared, he would have reached out and shaken me until I was a limp scarecrow.

"One does not reason with people like them," he said, his voice, brittle as the icicles that sometimes hung from the branches of the RishMorian trees. "The Crezands make use of intentional violence against the peace treaty. They choose to fight us. They do not wage honorable battle, but instead, they perform a Shymba war."

"What on Earth is a Shymba war?" I asked.

Shnize took another bite of his dinner. His eyes indicated that I should do so, too. I complied, but then I put down my utensil and asked again. "Tell me, Shnize. What does that mean?"

Shnize took one last bite. Then he pushed away his empty plate, patted his lean stomach, and leaned back. "It is a Franian term that the RishMore use. It comes before our earliest years of book history, when there is only one Queen — Crinana.

"Eat, if you want me to tell you the story," he demanded, like some male version of Scheherazade.

I shoved another bite into my mouth. The zepers had been to my liking. The bleo meat was not. It was merely something that Dar used to order frequently for me because of its reputation for making women more enthusiastic bed partners. (Not that he'd ever found me lacking. He had taught me well how to find my pleasure.)

"Crinana," Shnize continued, "they say woos many lovers, but she favors Vren. Vren is strong as the wood of the mena tree. He is smart as the storytellers of history, and he is handsome as all the males of his family. But one day when the Queen's lovers gather, Vren does not come on time. He is making passionate love with the Queen. The other lovers discuss this, and then they pool their evil and hatch a nefarious plan.

"The next morning when the Queen comes down to her table, Vren is not in his position of waiting. Crinana rants and buzzes her displeasure, but no one will say where her favorite lover is.

"It is three long weeks before the people find him, and by then, his body is full of decay.

"The Queen's guards carry the poor Franian's body to Crinana. See sees. She understands. His bones lie in shatters. In some places, they are like chalk dust. His face displays a mask of horror.

"The Queen weeps when she sees her lover's body. But she collects herself and orders him to the Death land where a lover's final rites occur. Then, wrapping him in the leaves of the straba tree, she buries her beloved Vren beneath the wide expanse of its overhang."

Shnize stopped and rose to take our dishes away.

"That's a terrible story," I cried out. "Why did you tell me that?" I asked him, wiping a tear from my eye.

"Because you ask me about a Shymba war. The story is not yet at its end."

"But the hero is dead. What more can you tell me?" I asked.

"Listen, High One, and you will understand."

I sighed, no longer the least bit sleepy. The meal had restored my energy, but still I leaned my head against the wall.

"The Queen goes into mourning after her lover is underground," Shnize continued. "Poor Crinana mourns him for seven and three, and then she wakes. The war begins that day."

"Seven and three what?" I asked.

"Seven days and three hours,' Shnize said, smiling because I think he has decided that I ask too many questions. But I can't help it. There is too much I don't understand about the story.

"What do you mean she woke? Did she sleep all that time?"

Shnize sighed. "She wakes from her mourning, High One. Throughout the seven and three she does not talk with a soul, nor eat nor engage in any pleasure."

"I wonder if I should be mourning like that?" I said, musingly.

"High One, this tale is very different from your situation," he told me. "The Queen, after she wakes, wages a Shymba war against the lovers who kill her Vren."

"I thought that's what you were explaining about, but I still don't understand what that means. What is a Shymba?" I asked, puzzled by his strange manner of telling the story.

He sighed again. Yet, his voice when he continued was as clear and calm as tranquil water, and he smiled at me.

"The Queen fights a battle that kills every man who murders her lover. She tortures the ones who are responsible for the disfigurement. She kills them all, and then, she goes to bed. In a week, she is dead."

"Why was she dead? Somebody retaliated against her?"

Again, Shnize smiled as if I'd just spoken the words, he'd wanted me to say. "She dies of grief. She is a weak Queen, High One."

"No, she wasn't," I argued, sitting up straighter. "She killed all the bad guys. That's not weak! She even fought a battle. How can you call her weak?"

Shnize's grin widened. "She is weak because she cannot endure her grief. She chooses to give into it. You will not. You will rise and be victorious against our enemies. Then, you will rule long and prosperously."

I still didn't understand about the Shymba, and I didn't agree that dying from a broken heart was a weakness, but my eyes were feeling heavy again. I'm sure I would have argued otherwise, but another yawn hit me.

"Enough talk," Shnize said. "You want me to undress you, High One?" he asked, not looking me in the eyes.

"No," I responded quickly, waking up enough to scramble off the bed. "I need to check…"

With the dishes in hand, Shnize turned and looked back at me. It was a look exactly like one Dar would have given me, a look that had nothing to do with a Franian's servile attitude.

He was not silent about it either. "I do not understand what a Prime Minister does," he said, "but as your private guard, it is my responsibility to see that you do not make yourself ill. Either order me away, High One, and I shall resign my position — or change into sleeping clothes — with or without my assistance so you can go to bed."

I wanted Shnize to be my friend, but I was uneasy about changing in front of him. I knew that was silly. Dar had continually lectured me about ignoring the presence of the Franians. *They are a door, a table, a rug. They are the picture on the wall," he used to say. "Do not notice them. They do not see you as a person either. They look upon you as the position you wear. They see only your status.*

I could still hear his voice saying those words. They were running through my mind. I could see Dar's taunting smile, challenging me, pushing me out of my Terran phobias, but it didn't matter. I couldn't shed my inhibitions.

Besides, Shnize had never been a door, a table, or a wall. He was Shnize, a Franian, certainly — anyone built like he was, a genie with a chest of steel — had to be Franian, but…

He was still staring at me, probably wondering what I was thinking. His eyes examined me, gazing into me as if he could actually see straight inside. For the first time, I saw a resemblance to Dar. Shnize displayed the same look I'd seen so often on Dar's face. Shnize was laughing at my shyness.

"I have decided to sleep in my current clothes," I told him briskly. Then I turned around and crawled beneath the bedding coverlet. Pulling it up about my shoulders, I lay there, staring up at him, daring him to say one word about my cowardice.

Wisely, he said nothing. He bowed his head briefly in acceptance of my decision, but his eyes glowed with suppressed humor. He shot a glance at me, one that covertly mocked me, but he was careful in his teasing.

"A ruler does not fear undressing in my presence," he said, as he deposited our dishes back into the food machine for reprocessing.

I kept quiet. I knew he was just simply stating a RishMorian fact, but what would he do if I actually said, "Sure, come undress me?" Would he really do that?

"Must I continually remind you that I am Franian?" Shnize asked, suddenly solemn about it, as he turned once more to look at me. "You do not worry that I might take advantage of your unclothed form, do you?"

It was a serious question, one meant to be answered truthfully. Yet, what could I say? Yes, I knew he was Franian. That was the easy part, but I also knew that Shnize didn't act like any other Franian I'd ever met.

Dar had said so, too. He'd warned me once about Shnize's irregular behavior. I couldn't remember exactly what he'd said. We'd argued about it, though. I'd thought at first that my husband was suspicious of him. I should have known that Dar would never fall into jealousy. He knew he was superior to everyone.

The room's heat was suddenly too great. I threw back the coverlet. "I do trust you, Shnize. I just can't do everything Dar did or everything he told me to do. I can't undress with you present. It would feel wrong."

Shnize walked over closer and stared down at me. "You will need to leave such feelings behind. Dar awakens you. He trains your body to certain needs. That you cannot ignore, High One. You will soon find that it is impossible to endure such marital deprivation."

"I don't want to hear this," I said, sitting up suddenly.

"Yes, but if you do not hear it from me, you will hear it from Zoff, your doctor. I do not know how sexual matters are handled on Earth, but you are no longer Terran. You are RishMorian. You will need a mate very soon."

"I'm warning you, Shnize. This is not a subject I want to talk about. Please, stop."

"In a moment, High One. You ask me to fill you in on details you may not know. This is one of them. I would not do my duty if I leave you with insufficient information. You will need to take a mate, of course, when we reach Crezand, but until then, your bed will be lonely."

I slid back down and pulled up the covers. Since Shnize wouldn't stop discussing such things, I decided to shut him out and …

"What!" I said, shooting back up again. "A mate on Crezand. You've got to be kidding. Me? No way. Fly eyes are absolutely not my cup of tea. There is no way that I would ever even think…"

"You are the High One, Prana. Of course, you will mate with the sacrificial offering."

"No, I won't."

"I will not argue with you, High One," he said, standing there as calmly as if we were discussing my favorite foods. He arms were crossed in a typical Franian manner, but his face wasn't representative of his race. His cheeks were tight, his lips turned down — it was as if he were almost showing emotion.

"Good, because marriage to a fly eye man is simply not going to happen!"

Shnize took a deep breath and let it out slowly. "We will discuss this another day. I realize you are tired tonight, but I also know you will find sleep difficult. Your body needs release."

"I said that's enough." I was angry then. I sat up and glared at him. He was pairing me up with a repulsive-looking mate and then telling me that I needed a sexual release? Who was he to talk? He was Franian, a null. He couldn't…

My silent musings didn't stop him. He continued on, flaming my face with even brighter heat.

"I have training in this, High One. Perhaps, I cannot bed you as a man does a woman, but I can ply your body with my hands and my mouth. I can fulfill your needs."

I covered my head with the coverlet. "Go away, Shnize. I told you I didn't want to hear it."

"You are still in shock from the death of your husband. I understand that, but when your body calls, when your brain is crying for love festing, I want you to understand that I can service you with my…"

"You will leave now," I said quietly but extremely firmly. "You could never replace Dar, no matter who or what you are, no matter how much training you've had in… in… just go!"

Shnize bowed, turned, and walked toward the door. Then he stopped. Without facing me, he continued. "I do not mean that I can replace him, High One, but my hands can assist you. It would help you sleep — even tonight — but no, I see, you are too angry for the thought of it. I will let the idea sit and brew. You will remember my words when the time arrives."

Chapter Three

I had not answered Shnize. I'd been rocked with shame — and a sudden flash of need for exactly what Shnize had suggested.

"Get out. Leave me alone," I'd yelled.

He'd turned again to look at me. Then he had bowed and departed, leaving me alone.

"Lights out," I whispered a few minutes later, but the darkness felt like a slap in the face. Shnize's final words kept replaying in my mind. I still saw the way his face had looked and the way his eyes had assessed my body. Oh, it was a doctor's look. I'm sure he had not craved me in a physical way, but his look had been filled with such sweetness that shivers were still sending electrical currents buzzing through my body.

It was then that the loneliness hit me even more fiercely. Waves and waves of it. I needed Dar. I needed his arms around me, the warmth of his body, the feel of his breath against my neck.

"Oh, Dar. I miss you," I called out, but he was silent. My alien lover, my husband, was gone forever. I sobbed great heaves of misery. I cried as if the tears of the past day had only been a prelude. An ocean of tears wet my pillow, and still the hollowness wouldn't let me stop.

"I hate you, Dar," I screamed. "I hate you for leaving me. Why? Why? You promised me. You said nothing could ever hurt you. You said we'd live together forever. You lied to me, Dar. You lied!"

Tears were the melody of my sleep. They played throughout the night and well into the morning, but my slumber was restless.

I hated the darkness and the length of the night. I hated the emptiness, the coldness of his side of the bed. I cried my pillow soggy, and when I dozed off, it was only for minutes. Each time I closed my eyes, I saw Dar looking at me, and his eyes were sad and hurt because he loved me, and I'd forced him to punish me again.

He was looking at me through his mind's eyes when the bomb hit. I could see him standing there, his Crezand fly-eyed wife, beside him, his hand pressing demandingly into the skin of her wrist. He was angry with her, and she with him. Waves of loathing filled the air about them. Then came the blast with the sound of a scream that never really happened. Yet in my mind I heard it. I felt Dar's eyes, the screams of both of them, the flash of pain, and I saw their bodies fly into a thousand particles of blood and bone and flesh.

It was a horror movie that replayed itself over and over — that moment before and then the actual occurrence. All night I witnessed the horrific scene in a loop.

I vowed not to sleep anymore, but then I'd drift off with fatigue, and each time the sight would be new and fresh, and the waves of pain didn't only emanate from Mada and Dar, but from me, for I was somehow there, part of that ghastly, gruesome moment. I watched Mada's hand twitch each time. I knew the import of her eyes, the hatred. I saw, and I wept, and then, it happened again and again.

I fought against it. I sat up. I tried to change the scene, to think of my boys, even a vision of long ago when I'd been a child on Earth, but each time, the picture repeated, and I was its unwilling observer.

Finally, in the wee hours of the morning, I plunged once more into the heaviness of sleep. I didn't want to go back to the dining room. It was Dar who dragged me there. But I knew that couldn't be. I knew even in my dream that he was dead, and I couldn't bear to watch his death again. But my mind tortured me, and once again I was a witness in the banquet hall to a scene that unraveled in slow motion.

Logically, the horror of it should have been dulled by my mind's frequent visits, but it wasn't. In fact, it seemed more vivid each occurrence as I saw details, I hadn't seen the time before. This time I even felt the explosion, and it was my skin ripping into shreds, my flesh scattering into a hundred bits and pieces, my blood spraying the hall with its crimson/burgundy stain. I woke up screaming.

Of course, the guards came darting into the room. Frick ran almost the whole distance to my bed before he halted and stared down at me. Our eyes met. He swallowed hard, seeing that I was both uninjured and in no danger. He attempted to speak, but he didn't know what to say.

I suppose a hysterical female, one who was also the High One, was beyond anything his training had prepared him for. He tried again but paused his tongue before he'd verbalized the thought. Then he sighed heavily, bowed his head, and said, "I will send for Shnize." He turned a military about face and ordered, "Retreat."

The guards backed out of the room, leaving me shaking and still as full of terror as I'd been a moment before.

My eyes were streaked from crying. I was writhing from the shock of pain and grief. I didn't know what to do. I just knew I couldn't sleep. Yet, I was so tired I could barely move. I swung my legs to the side of the bed and attempted to crawl out of bed.

I think I would have risen up and paced the room in another second. I was prepared to do that the rest of the night, but Shnize came rushing through the door, his eyes dark with worry, his face a mask of concern. There hadn't been time for Frick to alert him. How had Shnize known?

"What are you doing here?" I cried out, furious that he'd returned and angry because I needed him so desperately.

Shnize

It is not good that I leave her before. Yet, how can I stay? She orders me away. She is the High One. Of course, I obey. But staring into her eyes, I know it is a mistake. She needs me.

There is no rationale in what I do then. I am Franian. I do not have the ability to function under such conditions. I have no training in empathy, yet, instinctively, I open my arms and say, "Come to me."

She flies straight off the bed. I am ready for her. I gather her fragile, young body into my arms and treat her as I would one of the children. I hold her against my chest, my hand stroking her hair, and coo to her.

"It will get better, Prana. Soon you will be able to breathe again without it tearing up your insides. Soon, you will sleep without dreams, without the horror of your loss. One day you will even feel contentment. Life carries you out of this realm and into the next. Life improves, even beyond the After of someone you love. I promise."

She sobs against me, needing what I offer. She does so once, long ago, crying her woman's tears onto my chest. But that time I hold her stiffly, preventing her closeness, even pushing her away. I do not do that this time. I bathe her in the warmth of my caring.

My arms are tight around her small body. She snuggles up against me, seemingly as small as Frawn, yet, so different. The plumpness of her breasts attracts me beyond all telling, as do certain other features, which draw the blood to parts of me, which I do not wish to inflame. I move so she is not aware of it, but she is too lost in her grief even to notice.

Attempting to brace myself against the sweetness of her body, I analyze the sensation. I stroke her hair. It captivates me with its texture. It is supple. Its curly strands are chaotic and seem to interweave themselves teasingly about my long, heavy fingers. It is a fragrant part of her, too. I inhale the scent and long to smile. I wrap a lock about a calloused finger and shiver from the delight it gives me. It is wrong of me to think these things, yet I cannot throw my mind elsewhere. It locks inside the pleasure of this moment.

I bend over slightly to touch her hair with my lips. The fragrance of her body strikes at me again. My heart beats at a faster rate. For the first time, I want to crush the woman against me in a stronger grip, to cup her buttocks in my hands, to push her closer to the place where a second before I must move to push her away.

But I do not give into such impulses. I continue to croon to her, as to her sons, Frawn or Strandor. She is thinking at that level. She requires no more at this minute than a weeping child would need. Yet, her body is not a child's. That thought keeps pounding through the rhythm of my blood.

Her body is limp against mine. She has lost all purpose, all thought except her need for comfort. Her spine feels pliable and weak. I must build her up again. I sense that. I must bring back the sparkle of determination that usually lights up her eyes with animation. That is my job, not the dispelling of any hunger that I might feel because of my Queen-service-body responses.

"Let it out, Prana. Let go of your grief," I tell her softly. "You must feel no guilt. It is the fates. It is his time to leave us."

She shakes her head, but she is still too lifeless to argue — yet. I smile. How can I not? It is in her character to argue. It is the Prana I love.

She continues for several minutes, shaking her head, babbling with no meaning. She blames herself. I say nothing. I just hold her.

As she cries, though, my thoughts drift. Holding her, remembering why I cannot do more, my mind wonders back to that place in the cell of my childhood. The dirt-hewn corridors of my childhood hold tens of thousands of us, and I romp in the soft, padded dirt of its flooring.

My brothers play with me. Life is always good. We have food and companionship. We want for nothing.

But there is no mother's love. None of us have a mother, so we do not recognize the lack. Observing Prana with her sons teaches me about such things. It is from her that I learn the gentleness of child-rearing. It is from Prana I learn how to love.

I reach down and kiss her hair. She doesn't feel it. The softness of its texture on my lips drives me forward. I pull back her hair and dare to kiss an ear and then her neck. What am I doing? I jerk myself away and return to my thoughts about the hive. In the hive is the conditioning. I must recondition myself.

Yet my mind will not obey. It wanders here and there.

When I am five, the Reading occurs. I tremble with my brothers, listening for my name. A Queen's man calls out the decisions. We already know that few of us will stay in the hive. Jardon, my twin brother, the one I am closest to, they call. His eyes and mine meet. I bow. He is to service the Queen. He becomes a noble.

That means I cannot remain in the hive. Servers of the Queen, by law, must come from different families. But it doesn't matter. I have no expectation of staying in the soft, padded corridors of our hive. It is the destiny of all Franian males to go out into the worlds around us. We are a nation that knows that we long ago fail in battle. The RishMorians are our conquerors. To leave for other worlds is our duty. And as they order, always we Franians serve on planets not of our birth.

In the hive, the Queen produces too many. Most are eggs. But my brother and I are born. It is this that mandates our unique warrior status. Some lands prize our specialness. But all of us must depart in time. The RishMorians order this. They keep us in peace, no longer combatants except by assignment. It is the way of the world — change.

I hear my name that day. They order me to go to the RishMorian headquarters, the First Planet. I pale at the news. The very name of the enemy brings fear to my soul. Yet, I stand tall. I am a warrior by tradition, and my distinct heritage demands even more. I salute the hive and accept my assignment.

I face forward, not looking at my pseudo brothers that day. I contemplate my fate. I have no vision of a RishMorian. I hear that they feed on one's mind, robbing one of identity. Yet, I am chosen. I throw back my five-year old shoulders, raise my nose, and format my face. Let no one think a member of my cell lacks courage.

When my proper face is fixed, I smile. "Yes," I say to any of my brothers who are listening. "If one is to serve, it is best to serve a RishMorian master for HE rules the universe."

Several younger brothers cheer. I am suddenly their hero. I bow formally to them. Then I glance at my brothers of age.

Most of them pat me on the back, each of them looking slightly guilty from their relief that it is not they who must go to the planet of the RishMore. Only I of our cell group must travel to the dominant ones. My age brothers go to other outposts, planets with need of strong backs and loyal servitude. My brothers are content with that. It is the safer position, yet none of them say so. Such a thing is not suitable to mention.

Strarvor comments properly, "Our unit brings much wealth to the Queen."

We nod and smile among ourselves. To bring wealth to our hive is our function. A Queen's server, especially a royal child, to be the choice from our unit, brings glory. And then, a RishMorian transfer brings ten times what an outposter brings. Once more everyone claps me on the back.

However, one of my brothers, Cordaron, flashes yellow flecks in a Franian display of jealousy. "The RishMore eat you for breakfast, brother," he tells me half seriously as he pushes against my shoulder like an athlete building muscle.

The others laugh at Cordaron's words, but their eyes take on a look of doubt. Thus, when they clasp me for the final parting, their hugs are rough and absolute. Everything is always final after a Reading since there is never any return to a homeland. Such assignments are for life. But the special severity of my brothers is a message that they consider me on death walk.

I shrug and cast a snobbish look, one of superior placement, upon my face. My fist slams into Cordaron's shoulder with equal force as his earlier knuckle punch. "Smidgling," I call him in insult. "Your words are the spit of bitterness, for you are wrong. To be the choice of the RishMorians is a great honor. It pleases me. What could be better than to journey to the stars with the race that conquers all?"

Smirks appear on every face, but the last victorious words are mine. I degrade Cordaron well. The others give me parting smiles of approval as I leave to go to my assignment post. A ship loads those bound for the RishMore first. It is then I see Franians from different cells. Some of them are larger and older than I am. Many of them are surly. One fights his fate. They cart him off in disgrace. Another quickly replaces him.

Despite that, we are all onboard the RishMorian ship within the hour. Inside the vessel, we stand in a long line. At the end of it, soldiers stamp me with a tattoo that denotes my new owner's name and rank. My arm carries my assignment. Others receive nitrogen freezes on cheeks or foreheads.

The soldier who administers mine tells me that the information appears on a tablet with each drone's requirements, as well as the owner's preference for the location of the tattoo. I ask the soldier for information as to my new owner, but he does not know. The RishMorians are often secretive.

After the tattoo placement, which tears our eyes, yet we endure with no outward display of pain, we receive arm injections and must then lie down in flat sleepers. My outer mask reveals no excitement, but I fight the injected drug, hoping to learn something from the soldiers who transport us. But I hear nothing before our departure into space. I slumber in stasis for that first trip through the stars.

The sleep lasts several months since my planet is at some distance from the RishMorian capital. If I dream during that long snooze, I imagine it is with a childish expectation of adventure. In my visions, I am sure I don't worry much about the temperament of my new master. I know that he will probably take no notice of me for a good ten years. It is known that alien lords each have hundreds of Franians at their disposal. I assume that I will be just one more. I am wrong.

Dar's father, Demofian, meets the ship and deals personally with the disbursement of the newly arrived Franians. I think it odd for a kinglike personage to do so, but, of course, I do not question it. (I later learn that he likes doing so. Perhaps one might call it his hobby.)

When they wake me the morning of our arrival, I am silent, partly from the training they provide in my sleep, which teaches me the language of RishMore and its customs, but also because I feel deathly ill. The wake-up injection has not set well with me. I skip breakfast because of my nausea, and thus, weak from the long fast, I am barely able to walk.

That is why I am not surefooted as I journey down the exit ramp. I trip, and a quad leader of the High One, seeing my unstable condition, lifts me up and carries me on his shoulders. Frailty in the cell is cause for automatic rejection from the hive. This interference with my pride stains my skin a deep, dark gray. Demofian sees that.

"What is this? A sickly Franian?" he laughs since it is well known that my species is immune to almost every virus and bacteria known to the universe. "See that he is checked over," Demofian orders the quad leader. “Then bring him to me. I see from his tattoo that he is the one I ordered, a gift for the heir.”

So saying, the supreme leader of the RishMore, the High One, they call him, strides onward. I watch as the mobs of Franian guards, still unloading the Franian cargo, hide him from sight.

I recover quickly, of course. Land strengthens my blood and my backbone. A meal of RishMorian food, and I am soon vigorous again. I do not see Demofian for several days after that, but when I am settled into a schedule of daily weapons training and seminars on government and protocol, he arrives with his young heir in tow.

I bow low as is proper, but I am from a proud cell. I do not cower or tremble at the sight of the conquerors. Instead, I look them in the eye and assess them.

Dar laughs. “I thank you, Father, for this gift. You are right. This one is a proud warrior. I shall take to him well. The fawning of the others proves tiresome.”

“You are a unique and will be special here, as well as on your own planet,” the young boy tells me. “I shall see if I can train you into an officer. If so, that will please me well. Come.”

I am a unique but I am not special on my planet, but I do not say so. I set down the weapon I am working with and follow Dar out of the Franian’s military nursery. I never return.

Dar challenges me to weapons, to map reading, to arm wrestling. He wins, of course, in all of them. But he also challenges me to a game of Sepla, a Franian game that involves layers of thinking. I am the cell unit’s champion in my age group. I beat my new lord in four moves. That is a mistake. An angry RishMorian can take my life for such impertinence, but he speaks to me, discussing the game as if I am his equal. Not knowing any better, untrained as I am, I pour forth my knowledge. In the days that follow, Dar beats me, and then he wins and wins again.

After that, Dar does not play the game with me, nor does he ever invite my analysis again. I am once more his Franian servant, and he, the master. Yet, I think something alters at that point. I am only a child at the time, but he elevates me to First Guard, and he seeds me with Long Life. Thus, I remain at his side for centuries — until Prana comes. She changes everything.

I look down at the sobbing woman who leans into my arms. She is quieter now. Her tears are almost at the end, yet still, she clings and needs more comforting. Once more I quietly bend and kiss her hair. I know she cannot feel it, and I am careful to make no sound.

But she senses my movement. She gasps out another sob and then goes silent. I peek down at her face. Her eyes are shut, and her sweet face, although tears have streaked it with salty lines of grief, is smoothing out in repose. I listen a moment to the sound of her breathing. Her chest is moving up and down in a slower pattern as her breaths deepen. Once more, the feel of her breasts pressing against my chest makes my scepter grow taut.

It is a silly appendage, one that causes me only pain. It isn't required by a Franian guard. Doesn't it know that I can never utilize its sexual functions? Only a server of the Queen has need of one.

Other planets treasure Franians for that reason. We are safe as guards for women. Therefore, it is, in fact, a mystery to me why my scepter continually reacts to this girl-woman. I laugh at myself as I stroke her beautiful hair.

I hold her until my body sways with fatigue and my arms cramp. Then, I lift her up to carry her back to her bed. I have done the same thing with her sons many times, but with Prana it is far different. My need is a ferocious beast that growls and gnashes its teeth.

I moan when she awakens enough to circle her soft, dainty hands around my neck. Without pausing, clinging even tighter, she kisses my neck. "Dar, I am so tired," she murmurs softly, not refusing him, but merely expressing her feelings.

If I'd thought the pain of my need could grow no more brutal, I discover then how wrong I am. I could take the woman in her sleep-drugged state if only I were a full male. I could ride her as a man rides his lady, but I cannot. A Franian cannot. My history tells me that over and over.

With one hand, I flip back the upper blankets. Then I slide the sleepy, young body into the bed. But she does not let go of my neck. Her hands still cling.

"Dar," she says, "don't leave me. The fly woman can wait."

"I am not leaving," I tell her, expecting her eyes to open with the shock of hearing my voice instead of Dar's, but she does not react to it.

She only sighs. She does not fight me as I unclasp her hands. "I am so glad. Thank you, Dar," she whispers sleepily.

I start to cover her with the blankets, but I notice that her dress is sliding upwards around her thighs. I tug at the material, but it is a great ball underneath her. I know it must be uncomfortable. It might even cause her a bruise that she would blame me for. I lift her body up slightly so I can pull the fabric down."

She clasps my neck again. "I told you I was tired," she laughs softly. She lifts up and without opening her eyes, places a kiss on my lips. "But if it will keep you from the fly, I will…"

"No. Go to sleep, Prana," I say, once more, loosening her hold on me. "You need to sleep."

A smile tweaks its way across her lips, but she does not speak. She is already halfway into dreamland.

"Then hold me tight…" she mumbles.

Never is a man so unfairly the recipient of a temptation this great. I know I should leave her. It is wrong to lie beside her. But honor, duty, and obedience rule my life. I recline on the bed atop the blankets and take Prana into my arms, pulling her close, granting her a peaceful rest.

There is none for me, however. I watch over her, observing the rise and fall of her sweet, soft breasts, her delicate features, the soft pink of her beautiful lips — lips that have just touched my mouth for the first time and kissed me as if I were a RishMorian lord. I lie beside the High One perfectly still for many hours, and despite my heritage, I yearn.

Prana

When I opened my eyes, I started to scream. Shnize was beside me, his arms enclosing me, his massive body half-trapping me. Seeing that I'd awakened, he swung a huge, burly hand over my mouth.

"You are safe, High One," he told me. "Do not scream. It will merely upset the guards."

He rose away from me to sit up, but his hand still covered my mouth. I assessed the situation and tried to make sense of it. He was fully dressed, after all, and had been lying on top of the covers, yet I remembered that his arms had been holding me. Why?

But I was no longer panicked. I saw quickly that his eyes were not full of lust. He was staring down at me with deliberate patience and with the red-rimmed eyes of extreme fatigue. And besides, nothing would be gained by crying out. I nodded my head.

"Good," he said. "You are reasonable, High One." He rose up and stood beside me, bowing low. "In the night, you have need of my attention. I hold you while you sleep, High One. I only obey your command."

I remembered parts of that. I blushed, recalling how I'd run to him and cried in his arms.

"I do not need you now," I said sharply. "Leave me."

He bowed again. "As you wish, High One," he said. "For how long am I free?"

I wanted to tell him forever. I wanted him out of my sight as I recalled more and more of the night. "Did you carry me to bed?" I asked, unsure what was a dream and what was reality.

"Yes," he answered, standing straight, staring into my eyes with the expansive pride of his normal bearing.

"Did you – did you…?" I couldn't complete it. I moaned and turned to look at the wall.

"I only hold you so you can sleep. I am Franian, High One."

It must have been a dream where I'd kissed him. He couldn't be looking at me with such honesty, such unashamed integrity if there'd been more. I sighed.

"I'm sorry, Shnize. I was just surprised, is all — to find you in my chamber. Thank you for watching over me."

"It is my duty, High One. I am available for whatever you need."

I blushed again. Was he reminding me of his words the day before? "Return in an hour. I must bathe and get dressed. You also need to change. Your uniform is wrinkled."

I sounded harsh and crabby. I hadn't gotten enough sleep, but surely that was not his fault. "I'm sorry," I said. "I didn't mean it to sound like criticism."

"The High One need never give a lowly Franian an apology. I am at your service. You are free to yell or curse at me, or beat me if it pleases you," he said.

"Stop it, Shnize. I don't want to do any of that. Besides, I thought you were freed by Dar's death?"

"Not any longer," he said with a twinkle. "I am now the High One's Prime Man Star."

"Prime Minister," I said, correcting him. "Go, Shnize. Eat, too. We have a lot of work to do. Return quickly."

His eyes scanned my face with the look that he often gave when he was uncertain of a direction. I realized that I had given him a conflicting one. "Not quickly, Shnize. That was an error. Come back in an hour, please."

He bowed and stepped out. Then I turned and slid into the pool. Thirty-two laps wore me out. I was too tired to do more. I showered, washed my hair, and then dressed, wondering about the funeral customs for the wives of the High Ones. I scanned the computer and read several documentations of RishMorian customs, none of which were acceptable to me. I had no intention of slashing my face or cutting my hair. Nor did I intend to dye it white as a sign of my grief. I closed out the information box and began to pace, completely forgetting about breakfast.

Shnize knocked. The door slid open without his waiting for me to call out. It was a system that had worked for Dar, but I was not comfortable with it. I told Shnize that immediately and forbade him entering without my permission.

"I am a guard, High One," he responded. "I cannot wait for your spoken word. Suppose you are ill, suppose that a man has his hand over your mouth as I do this morning? I must be free to enter at will, High One. I cannot do as you request."

So much for having lots of power. The first thing I commanded Shnize, he gave excuses for why he couldn't obey me, but I didn't argue further. I had too much to do that day, and I was weary already.

"What should I be wearing to show my formal mourning for Dar?" I asked.

"You are the High One," he told me. "You may not mourn. You must act instead. That is the way a ruler shows his, I mean, her grief."

"But shouldn't I wear a black band or clothe myself in mourning colors?"

Shnize scanned my dress. As usual, I was wearing Dar's choice. Shnize sighed.

"I would have you choose a new color, High One. You must wean yourself from your husband. You must stand on your own to claim the respect of the planets we conquer." He passed me to look among my hanging dresses. "I would choose this," he said. "This is a color that suits you."

I groaned, eyeing the ruby-colored gown. "But I wore blue to honor Dar. Will it hurt to wear it just for today?"

"It is better if you do not. Yesterday, you wear Dar's colors. Today, you must stand alone," Shnize said, walking forward and shoving the dress into my hand.

I stood there holding it, thinking about changing, remembering how Dar had always picked out my gown for occasions of note. He had made me angry at times, but now I found I missed his tutelage. Dar! Even his name struck at my heart. I suppose the tears might have come in another second, but my stomach suddenly growled loudly, protesting its emptiness.

Shnize raised an eyebrow. "Before, you say feed my body. Do you not yet eat your after sleep meal?"

"I swam. That was better," I said, still eyeing the dress, trying to decide what color I wanted to wear for the rest of my life.

Shnize marched angrily over to the food machine and programmed a bar. "You must eat. Your strength is necessary for all of us," he ordered grumpily.

I nodded and agreed. "I'll go change first," I said, dashing away to the bathroom area.

When I returned, Shnize brought me a cup of the morning brew of the RishMore. I was thankful for that, but I nibbled the nutrition bar as well. I figured that it would probably look bad for a leader to have a grumbling stomach. Everyone was probably already laughing about the ignorance of the High One. I didn't need to offer them another joke for their ridicule.

"I don't suppose I could change colors, could I?" I asked, still stuck on something Dar would have said was superficial and unimportant. *Yes, Dar, I know*, I whispered in my mind, but sometimes, it is the superficial that allows one to go forward. Of course, I did not say that to Shnize. I knew he would have a ready answer.

"Change? Why do you want to do that? Your color represents your status. Once you declare it, no one else may wear it."

"That's silly, and I don't want to be stuck wearing the same color year after year."

Shnize bowed his head. "Forgive me High One, but I do not understand. The ship has damage. The chief officers are dead. We are returning to Crezand without a precise war plan, and the RishMore must formally declare your reign. That must take place immediately. Am I to understand this item takes precedence?"

It was a delicately phrased putdown. I sighed and then laughed. "Okay, Prime Minister, lead me forward. What is it I really should do first?"

And so, the day rushed by, one decision after another. Captain Jertot joined us in my personal chambers, something he had often done with Dar. We lunched from machine selections. None of us were enthusiastic about the nutrient bars we'd been living on for the past couple of days, but need was greater than time. But it did steer our conversation back to the dining hall cleanup and rearrangement.

By the time we were finished with that, or at least, our concentration was exhausted, we had covered all the issues. Most of the problems we had even come to an agreement over.

When the captain rose and said, "I need to get back to the ship's main post," I stood up also.

"I want to see the place where it happened," I said quietly.

"No," Shnize said, blurting it out with such force that both Jertot and I gazed at him with astonishment.

"She is the High One, remember, Shnize," Jertot said gently.

"She does not need to go there. Last night, she finds no sleep because of nightmares. Do you wish her to fall apart on us?"

"I am still standing, and I'm not falling apart," I said, grabbing one of Shnize's huge arms and uncrossing it.

He looked down at me with a strange expression, not fearful, and not oppositional either. "Prana, I cannot keep you from doing as you wish. As Jertot says, you are the High One. However, as your Prime Minister, it is my opinion that going there at this time would not assist you in your acclimation. It impedes it."

"So what am I supposed to do? Hide?" I asked him, irritated once more for no reason at all.

Jertot, meanwhile, I guess predicting a storm, took his leave and bowed out of the conversation. The door slid shut behind him before Shnize responded.

"It would be more beneficial for you to visit the crew members that are hurt. Do you forget them? Many have injuries and may die."

I sank down on the bed, my argument fatally ended. "I never thought about them. What kind of a leader am I? How could I have been so selfish? So unfeeling? So…"

"You are in shock, Prana. Yesterday, your husband leaves for the After. No one expects you to attend all matters at once. You are doing the best you can."

Shnize was standing there so stiffly, I was sure I could read what he wasn't saying. I had neglected my people. They should have been first.

"Oh, Shnize, of course, you're right. I must go right this moment. I'm so sorry. I'm an idiot. Why didn't you mention them? That's why I need you. I need you so much."

Shnize turned away. I thought he was angry with me, and I grabbed his arm once more to try to make amends. That was apparently a big mistake. It shattered his will. He twisted about and grabbed me in his arms. Then he lifted me up high in the air. "You will listen to me, little Prana. You will listen to me now. I will not allow you to be a Crinana. You will keep calm, and you will tread through this with decorum. Do you hear me?"

"Put me down, Shnize. I hear you," I said, not sure whether I should laugh or be angry with his behavior.

Shnize lowered me slowly. I stared at the ground, trying to understand why he'd reacted as he had. What had I done to cause the eruption?

He grinned down at me sheepishly. "You can order my death, High One, for doing that, or you can heed my advice. The decision is always yours."

I grinned. He knew he was safe. He knew I'd never order his death. "You would never have done that to Dar," I complained.

He returned my smile. "I would not need to do so. When alive, Dar is not emotional. The wounded would not worry him. He would terminate them."

"What!" I backed away. "No, he wouldn't. He would have…"

Shnize was looking at me with the wisdom of certain knowledge. I recognized the expression. "He didn't order your death," I said, still arguing even though I was pretty sure that Shnize had more than once witnessed what he was referring to.

"That is because of you, is it not?" Shnize prodded.

Then I remembered precisely how it had been. Shnize had taken my dose of poison, and Dar had not wanted to send him to the medic. I had begged and pleaded. Shnize was right. I had won his life back. Dar would have let him die.

"Yes," I said, finally admitting the truth. "But I didn't understand why. You saved my life, and he would have forfeited yours. I couldn't see why he wasn't more…"

I stopped. Dar was Dar. There was no point in analyzing it.

Shnize's eyes softened. "Many times, I thank you, but I do so now again. It is good to continue life. I am not ready to end it. Do you still wish to see those with injuries?"

I nodded. "Shnize, if you knew what Dar would have done, why didn't you advise me in that manner? Does it mean you didn't agree with him?"

Shnize grinned. It was becoming more of a habit as he loosened up with me. I liked it.

"You may still order their deaths. But, I think, it is not in your character."

I didn't bother responding to that. He knew my answer.

We set out for the medical lab soon after that. Frick wasn't on guard. It was his alternate, Staben, who marshaled the forces that made it impossible for me to see anything as we advanced through the corridors.

Staben was not one of my favorite guards. He reminded me too much of Stalt, the guard Dar had sent away. Shnize had told me that the two males were not alike in temperament, but simply from the same cell. Still, whenever Staben was on duty, I cringed.

Send him away, then, if he displeases you, Shnize mind sent.

I gasped and turned to stare at him. He had mind speech? Why had I not been told that?

Only those from my cell have the ability. We are of the Lineage that supplies the Queen's Drones.

"Did Dar know?"

No. Is it a mistake to reveal my secret to you? Will you inform the other RishMorians?

"No. Of course not. Your secret is safe with me. I am honored that you trust me. Can we talk about this later? There is so much I don't understand about Franians."

He flashed a look at the guards who'd formed a wall around us. They were watching the halls for threat, more than paying attention to us, but I suspected that at least Frick was listening in. Franians collected knowledge. Dar had once told me so.

So much to learn. But once again I was thankful for Shnize and for all that Dar had taught me.

Do I inform Frick that Staben needs replacing? Shnize repeated.

I kept forgetting I had the kind of power that would allow me to discard a guard and choose another. I turned the idea around and around in my mind. I closed my eyes to think. *Deep breaths. Leave off emotion*, Dar would have said.

Staben had done nothing to deserve disgrace. He did not set off my alarms. He only reminded me of Stalt. How could I treat Staben that way just because something about him…?

"No," I answered, rather uncertainly.

It's funny how much a mental smile feels like a kiss. My thoughts had pleased Shnize.

When I arrived at the public side of the clinic, Zoff, the medic, attempted to bar me from it. He was always too over-protective. I had to remind him of my new position. Then he paled, stepped back, and bowed low, allowing me full entrance. I hated doing that to him. I wanted him as a friend. Why couldn't there be a bridge between extremes? Why couldn't I have my way without demanding it?

While Zoff was still bowing and fawning, despite my request that he not, I looked about. The public clinic had been expanded by several rooms. The last time I'd seen it, (in fact, all the times I'd seen it) it had been empty. Maybe Dar had demanded it so, or perhaps, there had been other rooms I'd just never seen. I did know that on Dar's ship, rooms were often expanded and contracted as needed.

The next thing I noticed was Zoff's employment of four of the ship's robots. Robots weren't used that much since Franians were always available. Two of robots were sliding by on their multi-wheeled feet pods. I eyed them and stepped back.

They will not come in contact with you, Prana. Their program forbids that, Shnize informed me silently.

I nodded. Dar had told me that once before, but I still felt nervous around robots. We'd never had them at the Center.

One of the robots was carrying drinks and food to the patients. Another was distributing medicine. A third was rewrapping a bandage on a person I didn't know.

"Show me everything," I ordered Zoff.

"High One, forgive me, but as your doctor, I must speak. You are not up to this. Your face is still grief-stained, and it is obvious to me that you have not had adequate sleep."

"I will see all the patients," I repeated, not flinching from the lecture.

"Shnize, can't you talk some sense into her?" Zoff asked him, surprising us both.

My former guard simply shook his head. "She is the High One, Zoff. We must now do her bidding."

I almost laughed to hear him say that. But then I saw that he wasn't joking, as did Zoff, who promptly turned and escorted me through the clinic.

Many of the wounded displayed fear-filled eyes at seeing me. I spoke to them and attempted to relieve their worry by telling them that I would not penalize them for having been injured. I also added that they would be compensated for their pain and suffering, something that was not a custom with the RishMore. I ordered Shnize to keep a tally and dole out money from the royal coffers, which were plump enough to bear it.

Shnize said nothing throughout all that. I kept watching his face, but he gave me no indication whether I'd done well or had erred. It's funny how much I wanted to know his opinion. Dar would have been furious with me, both for the gesture and for my wanting a Franian's approval.

Yet there was sufficient gratification without my new Prime Minister's endorsement. Many of the men wept as I made my rounds, reassuring them that they would be healed not slaughtered.

"Thank you," they said with outpourings of gratefulness. But, apparently the policy of reward for injuries was foreign to them, and I saw that they viewed it with suspicion because I'd too radically changed their way of doing things. I realized again that I would need to go very slowly in making alterations to the traditions of RishMore. Even when it benefited the people, something new was to be feared.

At last I reached the final room. There resided the worst cases. Dar's former Chief of Staff was an empty cask of a person, hanging on only by artificial means. He would never be able to return to us.

"Disconnect him," I gave the order, although tears rolled down my cheeks — a display most unseemly for the supreme ruler of the RishMore.

"He would prefer death to this," I told both Zoff and Shnize.

Shnize nodded. *You have done well. It was the right decision.*

I continued then, viewing Dar's Chief of Weapons, his Head of Alien Affairs, and the Head of Internal Affairs. All three of them were stretched out on the medication beds, side by side, each on life support although for differing reasons. None of them, according to Zoff, had a good prognosis, but each had a chance. A new heart was being grown, a lung was ready, and a spinal column had already been replanted.

Dar would order their deaths, Shnize reminded me, meeting my eyes as a reminder of our earlier conversation.

I nodded and looked up at Zoff. "Give these men every opportunity to pull through — just as you've been doing. I want them to live. But if any of them reaches the same state as Grooger, end it."

"I obey, High One," Zoff said, bowing his head. "However, it is my belief that they may make it. I will do my best."

I nodded. Zoff was a good man. I trusted him.

"Thank you," I said. "You've been carrying a great load; I shall reward you as I have the others. What is it you would like?"

Zoff looked at me and smiled. "A woman, of course, High One."

That took me back, literally. I scooted into Shnize, not meaning to, not even conscious that I'd moved.

He doesn't mean you, High One, Shnize whispered into my mind.

Then what does he mean? I sent back.

Zoff had been watching our expressions. He laughed. Then he glanced over at the three men lying still and gray on the medical beds and gestured for us to move away.

I was ready to go. The sound of the machines' regurgitation and respiration, the room's sterile-clean odor which had a rather sharp, dry and irritatingly acetic smell to it, and the sickly hue of the three men I'd seen only a few days ago, cracking jokes with my husband, made me ill.

I suddenly felt nauseated. "I need to leave," I told Shnize. "I need to get out of here NOW."

"The High One will revisit your request in due time," Shnize informed Zoff. "However, she has pressing needs to attend to at the present time."

Zoff didn't question it. He bowed deeply, thanked me for coming, and led us to the door.

With Shnize propelling me forward, his arm in the center of my back, I exited. Then I breathed in deeply. It might not be fresh air, but it felt like it at that moment. I felt liberated from a great burden. I shut my eyes and tried not to see the faces of the injured. Why hadn't Dar been one of them? Why hadn't he lived?

I leaned against the wall. I knew my thoughts were irrational. Dar would not have wanted to live if he'd been damaged. Yet, I wished it. I wanted him there with me, whispering into my ear, smiling his superior smile.

"Lunchtime," Shnize said.

The dining room was, of course, out of commission, but there were many smaller dining places that were still usable. We went to a place called, "Threbmenti" which means more or less, "Quiet Stop."

It wasn't quiet that day. The place was packed. Probably around sixty members of the crew were getting meals. Some were standing and waiting for a place to sit. Others were eating at tables bursting with too many people. When the guards and I arrived, those waiting backed away.

"I won't go in front of them," I told Staben and Shnize.

Staben gaped. Shnize shook his head. "She doesn't mean that."

Shnize waved me to the side and said, "I suggest you reconsider. You need to give the crew confidence. Don't unnerve them by changing things. Pretend to be Dar for a little while. Put your nose in the air and walk into that room as if you own the room and everyone in it. You do, you know."

Shnize intimidated me even with his normal attitude. When he stuck his huge face into mine and towered down over me, I wanted to back up, but there was a wall behind me.

"Shnize, I won't be browbeaten," I said, staring up at him.

He reared back and gazed down at me. "Browbeaten?" he laughed. "By me?"

His amusement caused heads to turn. It wasn't enough that I had twenty guards strutting about the place drawing eyes; Shnize was making it worse. He'd thrown back his head and was making enough noise to frighten the whole Crezand planet before we even got there. I was sure they could hear him clear across the emptiness of space.

"Stop it," I cried out. "Everyone's looking."

That didn't make Shnize halt. He laughed even louder and fuller and made a big show about needing to prop himself up against the wall to keep from falling down. What was happening to him lately? I hardly recognized the quiet, introverted guard I'd once known.

I started to walk away. *Prana,* he said, and he reached out to pull me back, but his hand dropped before he'd touched me. Then his eyes traveled the room, aware as I was of the scores of people watching. Of course, he wouldn't want the crew to see that he, a mere Franian, took the kind of liberties with me that he'd been taking lately. I almost laughed, too.

"High One," he said again, curbing his impulse to turn me around to face him.

I stopped and waited. "It is not my purpose to bully you. I am powerless. I am only trying to do as you ask me. Isn't advising you of protocol what you order me to do?"

Why did he have to put it like that? I considered what he'd said about the crew wanting me to go first and about what he'd said about intimidation, and then I nodded hesitantly. There was no way around it. Shnize was right.

Thus, it was that I allowed the guards to elbow in, so to speak, and find me a seat. But when one was finally located, then Shnize refused to sit down with me. I wouldn't go that far in following his advice. "I order you to sit beside me," I said. "It's a requirement of the Prime Minister to do so."

Shnize sat down then, grumbling about my loss of face, but I decided to ignore that. If I were the High One, there must be *some* benefits. What pleasure would there be in sitting by myself with a thousand eyes focused on each mouthful? This way, at least, I gave them something else to talk and think about. Of course, that's not why I did it. Shnize, even when he irritated me, was still my only friend.

We had just received our food, and I was filling my mouth with the different flavors and deliciousness of it when Captain Jertot joined us. That was desirable. I liked the man very much. However, it became evident, even after he'd placed an order for lunch and was sitting there looking hungrily at the meals that Shnize and I were consuming, that he had come to discuss business. I told him to begin.

"The operatives in Crezand have just reported in as they do weekly. All is well. It was thought that there might be a spy hover on board, but apparently there has been no contact between the ship and the planet. Crezand does not, it would seem, know that we are headed back to wage war on them."

"What is a spy hover?" I asked, dabbing one of my vegetables in a sweet and sour sauce that was delectable.

"A bug," Shnize told me in Terran.

I gaped. Shnize was unbelievable. Was there no end to his knowledge? I'd thought he scarcely knew my home language.

The captain didn't speak. He didn't know what a bug was and tilted his head, questioning what had been said. Shnize explained before I could get a word in.

I was just about to ask about the spies when Shnize asked his own question. I gave him a look, but its effectiveness was destroyed by the arrival of Jertot's food. The captain was obviously very hungry. All talk stopped briefly as he crammed his face with a RishMorian style sandwich.

I pushed my plate aside and drank. The iced tea-like beverage tickled my throat deliciously. I emptied half the glass before setting it aside.

When conversation resumed, we went back to discussing the Crezands and what it meant if they were unaware of what their hostage/bride had done.

“It does make me reticent about blasting off any bombs,” I told the captain, which made him drop his fist full of veggies.

“We must call up the Minister of War then,” he said. “Immediately. Are you saying that you want us to…?”

“She’s not saying anything,” Shnize butted in again. “She’s fact finding and expressing questions.”

I glared at him. “Shnize, I wasn’t…”

“You wish to cancel the war?” he interrupted, his eyebrows raised up like exclamation points.

I glanced at the captain. He, too, had wrinkle lines across his forehead. I closed my eyes and thought about how Crezand had killed Dar. But it was Mada who’d done the horrific act, not her people. I tried to explain that.

“How do you think she obtains the bomb, High One?” Shnize threw at me. “Do you think she builds the weapon on our ship? Do you think she personally disguises it so successfully that all our technology cannot spot it? We scan her daily.”

“I understand that, but maybe it was only her government that instigated retaliation, not the citizens we’re talking about bombing.”

Shnize let out a RishMorian curse. I didn’t know what it meant, but I saw Jertot gasp and Staban, the chief guard, push forward to insert himself between Shnize’s chair and mine.

"Back away," Shnize growled at him, but Staban didn't move. He waited for my signal. I took a deep breath and nodded that everything was okay.

"Sorry," Shnize mumbled. "Prana, you must decide. Do you avenge your husband's death? Think of all those you see in the clinic. Think of the others who die. Forty-six of them, counting…"

"I know," I said. "I know."

"You must talk to the new Minister of War, High One. We must settle this. Battles must wage far in advance. They are like a building one must construct from the base up. Plans, materials, men to advance it — all the preparations have to be set into place. And then, once the attack starts, there can be no pulling back."

The two men were looking at me as if I'd committed the worst of blunders. I took another sip of my tea, sighed, and said, "Yes, all right. I do need to talk with the Minister of War. Can you set it up?"

I guess that was a dumb question. Jertot coughed and managed to hold in his laughter, but his elephant-like ears still jiggled, just like Dar's had whenever I'd amused him.

Shnize just rolled his eyes at me, looking almost human — well, if you didn't count the baldhead and the teal-blue skin. Franian eyes were really more humanoid looking than RishMorian eyes, so the gesture didn't look too strange, but no matter what Shnize did, he still always looked like a seven-foot-tall genie.

Jertot rushed in to explain. "High One, you command us all. The Minister of War will arrive at once if you order it. The entire ship waits only for your bidding, Great One.

Sure, I knew that — kind of, but I'd been on the other side of the equation for so many years, it was hard to remember that I now had the kind of power that Jertot was talking about.

"The Queen is inside her. The High One merely cloaks it temporarily," Shnize said. "When the battle flares, she will rise to the occasion."

It was nice to have a friend on my side. I gave Shnize a big smile, even if it was a rather doubtful one.

Jertot tapped his wrist and spoke into it. Most of what he said was uttered too quickly for me to understand, but I knew the recipient was Terriff, the RishMore's new War Minister.

"I won't know what to say to him," I whispered to Shnize. "I was never part of those meetings with Dar. I never wanted to be. I hate war."

Shnize shook his head. "The present is always a novel challenge, High One. Let Dar's history guide you. Let my words, Terriff's, and Jertot's blend the present with the past. Crezand *must* be punished. Otherwise, other worlds will revolt. You must remember that, High One."

I nodded. "I know. And I have to avenge Dar's death and the others who were killed, but I don't have to like it, do I?"

Shnize smiled. "No, my Queen, it is not necessary to like it."

Jertot clicked off the transcommunicater. "I told Terriff to meet us in the War Room," he said. "Does that please the High One?"

Since I had no idea what or where the War Room was, I just nodded. Thank goodness I had Shnize. Otherwise, I'd have been swimming around in a fog with no lighthouse.

Shnize indicated that I should stand. "I will escort the High One there shortly, he told Jertot. "She wants to visit her room first," he stated, just like we'd been discussing it before Jertot ended his conversation with Terriff.

I pretended that I knew what Shnize was talking about, accepted Jertot's courteous bow of farewell, and then turned to leave the eating room. But we hadn't gone for more than twenty feet away from the eyes of the diners, when I stopped and said, "Why are we going back to my room?"

"Keep walking. I will explain when we arrive," Shnize snapped at me, without the usual courtesy of my title.

Such mystery. I peeked over at Frick. He'd taken his place as chief guard again, his break over. He glanced at me, feeling my eyes. I smiled at him, but he didn't respond.

Frick was a tall, heavy blue genie, like Shnize, but, otherwise, he wasn't at anything like Shnize. He was cold, withdrawn, and sterile of personality.

"Did you have a nice break?" I asked the man, trying once more to connect.

Frick's eyes responded, growing dark and even more distanced. He examined me briefly, then his eyes centered on my face, not by meeting my eyes as Shnize would have done, but by staring at my forehead as if he couldn't bear to look into me and actually see a person inside.

"Yes, High One," Frick answered, robot-like. Then without another word, he dropped further behind me. It was very guard-like. Dar would have approved. He had never liked guards to stay in one place. Dar had preferred them to revolve around me, scanning all sides.

Meanwhile, I could tell that Shnize didn't approve of my talking to Frick. He'd stiffened, his body, as always, registering the constant barometer of his mind.

I sighed. I did that a lot. Dar had been working with me on curbing my facial reactions, and I was attempting to dull my projections, but it wasn't easy. I sighed again, this time more quietly.

When we arrived at my door, Shnize waved it open and stepped to the side to allow me to enter first. I did so and turned around. "Now what?" I asked.

"You must leave Frick alone. He needs to concentrate. It is inappropriate for you to engage him in conversation."

I nodded, not bothering to argue. Dar had always said the same thing.

Shnize must have been practicing the speech. Even the wording sounded like my husband's. But, it didn't really matter, anyway. Frick had the personality of a blank wall. Why would I want to talk with him?

I told Shnize that, thinking he'd laugh, but he didn't. His face hardened. "Franians feel, Prana. Never doubt that," he said, his eyes staring into mine with an almost frightening penetration.

I hadn't meant for Shnize to take the comment personally. I'd never considered Frick to be a cloned copy of Shnize. I tried to explain that, but Shnize gestured for me to stop.

"I must speak of other matters, Prana. Soon you will meet with your new War Minister. I think this Terriff is a dangerous man. I hear rumors that he talks about you. He may need termination."

"Termination?" I cried out.

"Yes," Shnize said, nodding as he hushed me. "When you meet him, be careful what you say. Show no doubt."

"How on Quibble am I supposed to do that?" I asked, feeling like I was standing under an avalanche of problems. "I'm full of self-doubt, Shnize. Of course, it's going to show. I don't want this stupid position, anyway, and Terriff is probably right to question my worthiness of it."

Shnize's jaw tightened, and his oval-shaped ears flapped heavily against his face. "You are Dar's selection, Prana. Do you doubt the wisdom of your husband?"

I turned away. My eyes filled with tears. "You'd be a better High One than I, Shnize. I know that. I'm sure you've thought it too. Apparently, so has Terriff. Have I ever met this new War Min…"

Shnize grabbed my shoulders and spun me around. "You will never say that again. Do you hear me? Never."

Once more, he'd lifted me off the ground with the strength of his hands. My legs dangled like limp strings.

"Put me down, Shnize," I demanded, but with a rather uncertain voice.

Shnize stared up at me. The pupils in his eyes had doubled in size. He shook his head. Then realizing what he'd done, he lowered me to the ground. "Forgive me, High One. I did not mean to…"

Unfortunately, he'd still not released my shoulders. His fingers pressed into my skin like I was modeling clay. "Let go, Shnize. You're hurting me," I interrupted.

I thought that would end the situation, but it didn't. Shnize stood frozen in place with such an expression of horror on his face, it was if he'd aged a hundred years. Then he dropped his hands down to his sides and flung himself down on the floor. There, he banged his head on the carpet and cried out. "Your words of before disturb me, High One, I beg you to release me. I cannot continue as your Primal Minister. I cannot."

"Prime Minister," I mumbled, correcting him, at the same time trying to make sense of his reaction. What had I said that had fired him up? Why had he lost control? I replayed my words.

I shook my head. "No," I told Shnize. "Now get up. I won't release you. Not ever."

For once, Shnize didn't respond to a direct command. He remained on the floor, banging his head, groveling at my feet, and wailing like a mad man.

"Shnize, I command it. Do you hear me? Get up. I need you. I can't do this without you. Now get up!"

He banged his head one final time, let out a loud and plaintive half sigh, half moan, and then, very slowly, rose, his eyes great pools of horror.

I reached out and touched his arm. "I'm lost, Shnize. I'm sinking. I want to go home, but I don't have one. I can't go back to the palace. That would be the worst. Dar is there in every path and chamber. And I can't go back to Earth. I have two alien children. Besides, I'm not sure I'm still Terran. Am I?"

"You still do not understand," Shnize burst in, his eyes a blaze of such furious lights I backed away.

His voice, sharp as a razor's edge, cut into my panicked words and halted their flow. He'd simply restated what I'd been trying to say. I burst into tears. "Yes, Shnize, that's exactly what I mean. I don't understand. I don't understand any of this. Dar sheltered me from it. He didn't trust me with this kind of information. He knew I wasn't capable. He knew I was just a…"

"… the High One of the RishMore, the Queen Mother of the Universe, the decision maker of all policies, the war lord of a thousand kingdoms…" Shnize interrupted and finished my sentence.

Then again, he grabbed my shoulders, and this time, he shook me. "You are correct, Prana. You do not understand. You are the High One. There is no going back. No Terra. No returning to First Wife. Don't you see, Prana? There is only one release. Death."

That was a sock in the belly. I had no sighs left. I flung myself at Shnize, grabbed at his neck and hung there, crying against his chest.

He held me without speaking, his hands caressing my hair. His arms rocked me back and forth, and he hummed, just as Dar had done so many times.

I used him. I soaked up his compassion, his understanding, and the firm solidity of his heavily muscled body. Then it hit me. I felt more comfortable in the arms of my former guard than I'd ever felt in Dar's. That shocked me. But the truth is that Shnize cared. He commiserated. Dar never had. He'd just waited for his will to be absorbed.

"Prana, stop it. Ease it off now," Shnize said as if he were following my thoughts.

"You must practice," he told me. "You must grow the 'tiger.' Isn't that what Dar tells you? I forget what the word means, but you must find it inside you. You must grow the tiger."

I wiped my eyes with my hand as I attempted to restore my courage. My sobs stopped. I concentrated on feeling brave.

"I do not yet explain the worst, Prana," Shnize said after a minute.

I looked up at him, eyeing his large ears, his blue-tinged skin. "It keeps getting worse every day. Doesn't it? Please, I don't want to hear anything more that's bad," I begged him.

"The boys," he said, and then he was silent.

That swallowed up my tears. "What? What about my sons? Aren't they okay?" I asked, pulling away and searching his eyes for understanding.

"Yes. They are safe now. Frawn and Strandor are in no danger yet, but I double the guards on them."

"Yet? What do you mean, not yet? I must go to them…"

Shnize restrained me. "No. They are safe while we meet with the War Minister. But they are in danger unless you can find the tiger."

"What are you talking about?" I demanded, furious with him for stalling.

"Should Terriff continue to talk ... Should others have doubts about you..."

This time, it was I who seized Shnize's neck in my hands. It was a reach up, but if I could have, I would have shaken him exactly as he'd shaken me. "Stop it, Shnize. Explain," I said, hearing the sound of Dar's voice in my command.

Shnize nodded. "Yes. That is better. You are finding the tiger inside you. I see now. The tiger connects to your sons. I understand."

"Shnize, you are the most frustrating..."

"Think about it, Prana. If you are not the High One — if such a thing is possible, which it cannot be if you live — there is no one to protect the sons of Dar. A new leader would terminate them."

"Terminate? Kill my sons? But..."

I was slow to understand his meaning, but with a good reason. The politics of alien kingdoms was not in my knowledge bank. I didn't think like that. I thought like a Terran, one who'd never even known a supreme ruler.

"If Terriff is dangerous to my sons, I want him killed," I said. "Now."

Shnize chuckled. "The tiger wakens."

I took a deep breath, and then I flared at him. "Did you hear me, Shnize? I want him dead. If he plans to kill my children, I want him dead."

"Good. If he makes a move, Prana, I will see that it's done.

Shnize cleared his throat in a very unFranian manner. It brought my attention to his eyes. Were those tears?

"I cannot sire children, Prana, but I feel like the sons of Dar are my own."

"Thank you," I said, reaching out again and touching his arm. He didn't back away like he usually did. Instead, he bowed his head.

"I'm glad you feel that way, Shnize," I continued. "The boys need you, just like I do."

Shnize's arms stayed at his sides, but the place where I rested my hand on his chest, spasmed. I didn't move my hand. I needed that contact.

"I will protect the boys AND you, Prana. You have my oath on that," he said with a look that sent chills down my spine.

"Come, we will not mention this conversation again. Let the War Minister plot. We will pull him down. Then, you will be respected as a worthy adversary. It will lead others to caution."

As if he hadn't been speaking with such urgency, I found that Shnize had walked us toward the door. He waved it open, ready to leave.

"Wait," I said suddenly. "I have to go…"

I didn't finish my sentence. I disconnected my hand from Shnize's arm and dashed to the bathroom. The RishMorians and the Franians have kidneys much larger than mine. They could go all day without using the toilet, but I couldn't. Sometimes I almost forgot that.

Chapter Four

The new War Minister, Terriff, was even larger than Shnize. His body had a more muscled look than those of the other RishMorians. Yet, there was no question what species he was. The moment I entered the room, his mind assaulted me, testing, probing — a RishMorian noble of the highest rank. I thrust back with equal measure, glad for the training that Dar had given me.

Terriff's hands flew to his head, and he took a step backward in surprise. I didn't comment, but I smiled. It was enough. He knew.

Captain Jertot introduced the man to me formally. Such rituals seemed like a waste of time, but I waited it out, hearing the long list of his heritage and the successes of his career.

"Fine," I said coldly when it was finished. "You sound like you *might* be qualified for the position of my War Lord."

Shnize snorted — his form of a chuckle. I ignored it, but it drew Terriff's eyes. "Out," he ordered Shnize. "Surely you will not allow a *Franian* to hear our plans," he said, looking at me.

"Shnize stays," I said, my nose hoisted higher than Terriff's, at least it would have been if he hadn't been such a giant of a RishMorian.

"At least order him to stand guard at the door," Terriff grumbled in irritation.

"You dare to command the High One?" I flared, and Terriff lowered his head in submission.

Shnize

She is brazen and strong. She is everything Dar tells me. I have doubts in the past, but no more. Beyond her softness is the core of her. Now she flies into her new realm. Look how the War Lord grovels at her feet. She does not give him time to think. Good girl. Hold him to the defensive.

She gives him my new label. His eyes twist up with a look that tells me I am right to suspect him. A true believer does not question the High One's right to rule. Terriff accepts her every decision overtly, but he glances at me and smirks, holding on to his secret thoughts.

I stand silently, ready to play whatever role Prana wants of me. I do not intrude. It is best if Terriff sees her strength.

The War Minister lays his plans on the table. He speaks of the intended attack, positions of insecurities, and the weaknesses in their civil defense. Prana's eyes watch him as much as she attends to the plans. She is attentive, yet she does not comment. He grows nervous with her stillness.

"Have I bored you, High One?" he asks with a look of concern, but with the slightest hint of a certain smugness.

"No, I am listening," she says, calmly. Yet I see the slight narrowing of her eyes. He is unaware of it. He does not know her, as I do. She is far from bored, and his insult enrages her.

"Tell me about the bombs,' she says. "How many people will be killed?"

A curl of Prana's hair falls forward. Terriff is not immune to her charm. His eyes stray. They take in the gentle curves that the soft fabric of her dress displays. Anger quickens in me. I know his thoughts. The fact that he thinks them of the Highest One is further proof of my suspicions.

"Terriff," she calls him back. He looks up and sees her alien eyes. Then he is lost. Prana's eyes are deep and mysterious. Dar once calls them mud-colored, but that is not true because they shift with her thoughts and with the color of her mood. Her thick eye lashes, so unlike Franians,' which are nearly hairless, are similar to the RishMorians' lashes. They enhance the depth of her eyes. That fact suddenly catches Terriff's attention.

RishMorians find alien women exotic and exciting. Mysterious eyes often incite a special wonder. Prana's appearance is spellbinding, Dar tells me once.

"Terriff," Prana says again, and the War Minister shakes his head as if to direct his mind back to the plans.

"Ah, yes," he says, and he steels himself to answer her, but his eyes are traveling her body once more, marveling at her great beauty, considering how he can take advantage of it.

Prana does nothing to attract his fascination, but still, she enchants him. I see it in his eyes as he plots.

Terriff speaks the ranges of each type of bomb. Prana listens with hardly a blink. Her sharp mind is analyzing, seeing casualties, studying facts and results, churning the possibilities in a way that Dar attempts to train her, but thinks he fails. (Dar tells me one day, a little over a month ago when he drops by to see his sons, that he suspects his wife is not mature enough to learn war discipline. It doesn't upset him greatly, but he sighs, most unlike himself. Dar finds the delay of a goal to be tiresome).

If only he lives now to see Prana. She rises to maturity in her need of it.

Prana is still analyzing and questioning when I turn my attention back to the conversation. However, I cannot concentrate on what she is asking. Terriff moves too close to her. He is almost touching my

Queen. My fingers cram into fists. I shoot a glance at Jertot, but he does not notice. He listens intently to the discussion.

"High One," I say to draw her attention. She looks up and tilts her head slightly in the way she has of questioning someone. Her eyes as they glance at me are pensive. She temporarily forgets her mistrust of Terriff with her concentration on the battle maneuvers.

"As you ask me before, I now remind you about visiting your sons," I say, hoping the mention of Frawn and Stalt will remind her about Terriff.

"Ah, yes, Dar's sons. Do they have their mother's beautiful eyes?" Terriff asks.

I startle her. She turns a confused look in my direction and then glances at Terriff. "My eyes?" she says, flustered by his question. "Dar's sons have RishMorian eyes, of course," she says, taking his question as an attack.

"Forgive me. I did not mean that as it sounded," Terriff says apologetically, reaching out to touch her arm.

I erupt. Three strides, and I am upon him. "To touch the High One is an offence. She is far above you." I say as I stand with a poisoned knife at his throat, awaiting my Queen's command. Unhappily, she does not give it. Instead, she orders me to back away.

"You see," she says. "I am well taken care of. My Prime Minister is quite handy with a weapon, wouldn't you say, Minister of War?"

Prana rises then and covers a yawn with her hand. Terriff's eyes are once more pleasuring himself with a scan of her body. He focuses on the delicate mounds of her lovely breasts. I growl silently, seeing his eyes harden, his ears curl with desire.

"Come," Prana gestures to me. "It is indeed time to visit my sons. Thank you, Terriff and Jertot, for your time."

She nods her head regally at them, and they bow low, but as I back from the room, my eyes meet those of Terriff's. He glares dagger thrusts. I meet his look with an equal measure. So, thus, do two enemies mark their turf: I to protect, and he to conquer.

"I want to go back to my room first," she tells me as we leave the War Room. I am glad of that because while it's still fresh in my mind, I have several things I wish to speak to her about.

But as we enter, she waves the door shut before I can do so, throws her arms around my neck, and raises herself on tiptoe to kiss me on the cheek. "You were wonderful," she tells me.

I stiffen. Her behavior is entirely inappropriate. She must not touch me. She needs to keep her distance. Yet, I do not move away. I stand rigid, as inflexible as a wall.

The night before, I offer to fulfill the High One's needs. I do not forget that. But this is different. That is my touching HER body, my granting her release. There is no sharing, no reciprocation. She refuses my offer. She spurns me, indicating that I am unworthy.

It confuses me now. What does she wish from me? What does she require me to do? I do not know the answer, nor do I understand her moods.

Dar is everywhere in the room. I feel his presence. It is because of him that I do not respond in an unsuitable manner, the one which my body urges. Still, my heart races, beating at speeds abnormal to it. My breath becomes raspy, my muscles twinge and bunch in frustration, yet my arms remain at my sides.

I force my thoughts to center on her elevated position as the High One. She owns me. I remind myself of that, although it is not precisely true. Yet, if she orders my death, the guards cannot question it. Only later comes the need for restitution. She has power over me. She almost owns me. The difference is only phraseology.

I repeat this over and over as she keeps her hands in position, her chest against mine, her breath on my neck. She is the Supreme Ruler of the RishMore. Rigidity of body is all I can do. Breathe in. Slow the racing of my blood. Exhale. Restrain my arms. Do not allow them to cup the softness of her disturbing swells. Do not move forward, although I yearn to feel their heady softness more intimately against my straining pectorals. My scepter throbs with the frustrated ache of her closeness.

Prana is unaware. She talks about how bravely I came to her aid, her lips inches from my face. I wish to lift her up, to touch her lips to mine, to mold her body to mine. Her softness fits so perfectly.

I do not understand her speech. I blink and attempt to concentrate on the words. I imagine I am not breathing in the smell of her, the sweet nectar of her pores, the honey of her hair, the taste that lingers in my mouth, and the flavor I can sense her body gives out if I only bend to tongue her skin.

She speaks of how I use my knife to save her. She does not know how ready I am to take the noble's life. I would slit the man's throat in a minute with her permission. I lick my lips at the thought of it.

Terriff's eyes show lust for my Queen. His sweat reeks of desire, sending droplets all about the room. I yearn to halt his desire. Yet, I know that is not possible. I cannot strike without her word, and she has no violence in her soul, only purity and trust.

Of course, I defend her. It is my vow to protect her always. I tell her so, but it does not quiet her flow of gratitude. She praises me in that same vein for several more moments as I stand in the torture of her bittersweet web, her arms not around my neck anymore, but still touching me, soft and warm as her body. I sigh. In my ears, my voice, responding to her words, holds great sadness, but she does not hear that either.

Then, like the winds on RishMore, she blows a cold gust across my ache of desire. She plunges into a different topic, one that causes the sudden frost inside me to melt and then churn and boil. The heavy winds of rage sweep across my thawed heart.

"What do you think of Terriff?" she wants to know, her face filled with excitement and an eagerness I have not seen since Dar's death. "I handled him well, don't you think? Do you really still believe he's guilty? He didn't seem like it. He seemed eager to share his war plans with me."

She is such an innocent. How is that possible? How is it that the demands of Dar's training cause no theft of her trust?

Then I remember a conversation. Dar's lips, sloping upward at the corners, his eyes softer than they normally are, as he speaks of his amusement with his young wife. She says something during a discussion, uses some naïve expression of her sheltered inexperience, and he barely keeps from laughing, but he smooths it over and tends to her ruffled pride.

Her suggestion is suitably ignored. In the High One's retelling of the story, there is no mockery in his eyes. He savors the sweetness of his wife, her ideas of right and wrong, her innocent nature.

The memory pains me. At the time, I feel anger that Dar recounts Prana's ignorance, then describes her widened eyes when she speaks like an enthusiastic child.

"Prana is not a child," I want to say. But, of course, I do not. Dar continues with his thoughts, praising Prana's skills at analysis. I feel the urge to mention her drive to learn the RishMorian language and its culture, and her ability to do so with unparalleled rapidity. She is an amazingly quick learner, much faster than any other wife of his.

Most of all, I yearn to mention to Dar how she is capable of soothing and calming an irritated ruler with a diplomacy far beyond the beauty of her face and body, but of course, I say none of that.

Dar is not talking to me then, anyway. He merely uses me as a sounding board, listening to his own thoughts as he views his wife through distanced eyes. I am sure he leaves to meld with her again, to celebrate his great victory in finding the mate of his vision.

But the memory of that dialogue awakens new understanding as I realize suddenly that Dar has purposefully left Prana innocent and full of trust. I can understand that. Who would want to destroy such gentle sweetness? I do not wish it so. Yet, it leaves her vulnerable now. How can she be the High One and continue as she is?

She waits for my answer. Her eyes are impatient, but she does not interrupt my thoughts. Dar teaches her that as well. She knows to watch eyes and delay speaking. Once again, I witness her potential. None of his other wives gain such insight, although tutors teach the same lessons.

I turn and watch as Prana moves away from me. She walks with a royal stride, her head elevated, her arms tightly controlled, her body a fluid movement. Dar cannot train away the dance of her body movement. It is the way of the Terrans, I discover, watching Prana's Terran guard. Even that woman, military trained, carries traces of dance in her steps.

Prana sits down on her bed. Her eyes do not raise to look at me. She is unaware that I observe such simple things. Thus, I see the flash of leg as she pulls up her skirt so she can sit in the strangely conjoined position she prefers. Dar corrects her many times for sitting that way, legs wrapped in legs. But it is one of the things he at last accepts about her as they join their lives together. Once, I even see him smile fondly as she reformats herself into her favorite position.

Before Prana, never does he modify an opinion. It is one of the unique qualities of Prana. Even Dar softens under the sweetness of her manner.

She tugs at her dress to cover her legs. She is modest. I am grateful for that, yet her position still disturbs me. I look away.

"You handle him most appropriately, High One," I finally say, smiling at her. "He feels the insults and stings from your sharp jabs. He observes your ability in equal measures. Thus, you enchant him."

"What?" she says, blinking because my words confuse her.

I ignore her look and continue. "Terriff is very guilty, my Queen. His eyes are full of it. Do not trust him. Be ever cautious in his presence."

"But …"

Her eyes darken with the pain of her mistake. I hurt her with my words. I do not mean to rob her confidence.

"As I say, you behave in perfection. Both Jertot and Terriff find you impressive. But do not trust his change of mind," I say, seeking a delicate way to explain.

"He changed his mind? You mean he's not going to challenge me after all?" She cries out, her face open with her relief. I worry that she forgets Dar's training with such an openness of expression.

I sigh, sit down in a chair, and think about this new obstacle. Her face lights up with smiles. She looks happy. Since Dar's murder, that facial expression is absent. I sigh again, something I do much of late.

"No," I say with as much gentleness as I possess. "I do not explain well. Forgive me."

Her happiness dims. She watches me, her eyes forming her mask again. That is good, yet I vow to myself to find her another tutor. Her old one dies in the bombing. She will need to begin again, this time with much more urgency. She must not display her thoughts with such outward gestures and facial motions.

"Terriff changes his mind about killing you," I tell her, seeing the puzzled expression in her eyes escape without her awareness. "He decides to take you as his …"

I search my mind for a word to say. I do not know what 'whore' is in RishMorian. Even if I do, I cannot say that to Prana. She is too delicate, too fragile for such knowledge. I embellish. "I do not think he intends marriage."

Prana jumps up. "What? What are you talking about? He didn't say anything about marrying me. He wouldn't dare suggest…"

Then she stops, hearing my negative. "He DOESN'T intend marriage? I don't understand. What do you mean?"

I want to leave it as it is, but her eyes are demanding. I sigh again. "Maybe he thinks of marriage," I compromise for the sake of her innocence, "but I doubt it. There is nothing to gain by marriage to you."

She clamps her mouth shut. Then she probes my face, trying to put the pieces together. She walks closer and stops, standing above me, looking down. Apparently, she comes up with something close enough to the real meaning. Her face clouds with anger.

"That's disgusting. I would never…"

"If he overthrows you, Prana, it is a suitable method of showing superiority. Your death is safer and easier, but keeping you alive shows that he spurns your retaliation."

"He thinks I'm weak?"

"Yes. I hear he calls you the 'play toy of the High One' " I tell her.

Her nose flares in sudden anger. Maddeningly, I find it appealing. My scepter throbs demandingly.

"Play toy? I bore the High One two sons. Is that not worthy in RishMorian eyes?"

I nod, my eyes watching her breasts. They, also carry her anger. They jiggle delightfully, bobbing in front of me, in the rhythm of her faster intake of air.

"Even if he were to ask me to marry him, I would not do so," she says, her lips pouting.

Their enlargement reminds me of the way she looks after her kisses with Dar, and especially, after an afternoon of lovemaking.

I do not speak. Her words don't follow what I am imparting. As I stare at her, she blushes.

"All right. I understood what you're saying, Shnize. I'm not a total idiot, but I'd kill myself before I'd let him take me like that. I just meant that even if he decided to marry me, I wouldn't. I'm never marrying anyone again. Ever."

I do not mention that suicide is not an option, not with the long life serum Dar gives her. She must know that. I pounce on the more important part of her words. "Never say that again, High One. I beg of you. I understand your pain. You feel the loss of your husband. I know your grief, but you must not say those words again. It is your duty to marry and produce heirs. You have no choice in the matter."

"I did that. I produced two royal heirs!" She grabs a pillow and throws it against the wall. I watch the small, round head holder that she uses bounce off and slide down. I do not understand my Queen's action, but I suppose it gives her some relief.

She stamps off in the direction of the bathroom. She is normally a quiet walker. This, too, must be a frustration reliever. If so, it is better than making false statements.

For a moment, I ponder her frequent use of the toilet. Perhaps, I need to discuss that with Zoff, but then I remember that such behavior isn't new for her, and all the Terran guards when on the ship share this same tendency. I suppose their bodies are so small they cannot store water for long periods as we do. I make a mental note of the fact so I can better serve her needs in the future.

When the High One comes back out, she is much calmer. I do not bring up marriage or Terriff again. I wait to hear her wishes.

Chapter Five

Prana

I vowed to myself not to lose my temper with Shnize again. I didn't like who I was becoming, and my behavior was ridiculous. I was like the person who asked for news and then yelled at the carrier for telling it to him. Poor Shnize. He was so patient with me.

"I'm sorry," I told him. "I'll try harder to listen to you and not to argue all the time."

He blinked and looked at me strangely. "I do not understand," he said. "For what are you sorry? It is your right to argue. You must do so as our ruler."

"I didn't have to erupt at you," I told him, looking up at the ceiling so I didn't start crying again. "After all, it's not your fault what the RishMore do. I know the leaders have been practicing that tradition for millennia."

Shnize's mouth opened into a small, perfect circle. Then he burst out with a laugh. "You find that you like the idea of sex with many husbands, then?"

"No!" I yelled, horrified by his misunderstanding. I stamped my foot in my distress, a habit I hadn't resorted to in years. "No, that's NOT what I was trying to say! Don't be horrid!"

I looked around for my pillow. When I found it, I picked it up and hugged it against my body. It would be nice to use it as a punching bag for my frustration over the mess Dar had left me in, but I resisted that, turning toward the bed and placing the soft, blue pillow that Dar had given me back in the head position of my bed.

"There's no way I can do that. Do you hear me, Shnize? I'll change the laws. I'll put a stop to the practice. I just can't..."

Shnize's mouth closed. He scratched one eyelid, a sure sign that he was having difficulty knowing what to say to me. "Of course, I hear you, Prana. I just do not understand you. Why do you bring this up again if the subject displeases you?"

I stared at him for about fifteen seconds. My mouth dropped open to give a ready answer, but I stopped and thought about his question. I couldn't think of a good reason, or how we'd gotten started discussing the subject again. I closed my eyes and tried to remember what we'd been discussing before it came up. Terriff. We'd been arguing about him.

The image of the man flashed before me. He was a typical RishMorian — blue-skinned, diamond-eyed, bald, but he'd had an irresistible smile, and he'd flaunted that while he was sending thoughts at me, those stray flickerings of teasing compliments meant to tickle my brain with interest. I'd thought that he was doing it because he liked me. Yet all the time, he was just thinking about... An insidious idea sprang into my mind. I let it out and into the room.

"I've got a great plan," I said, laughing. "Here's what I'll do. I shall declare that henceforth all alien species will mate with my Minister of War. That will be part of *his* job," I said, giggling.

I thought it was a clever joke, but Shnize didn't see the humor. His eyes clouded up with dismay, and he shook his head with emphatic urgency. "No," he groaned. "That is wrong, High One. You cannot do that. Do you not see that if the War Minister services the alien wives, he is fulfilling the job of the High One? That is one of the primary functions of a RishMorian ruler."

"Better him than me," I said, trying to tease Shnize out of his seriousness.

Instead of laughing with me, Shnize grabbed at my shoulders. "Prana, for some reason you find this humorous. You cannot mock such responsibilities. This is serious. You must handle your obligations with honor and gravity and never make a joke of them."

"But it is a joke. The whole concept of conquering the universe is crazy. Why do the RishMorians need to subjugate everyone, anyway? Why can't they just let other planets exist and be content to ignore them? Why do they have to interfere and give me fly-eyed males as bed partners?"

"It is not *they*, Prana. You are a RishMorian now. It is you who rules YOUR people."

Shnize's words brought back my tears. I couldn't even jest about things anymore, and I was trapped in the horror of it. I wiped at my eyes. That didn't stop the flow.

Shnize walked over to the bureau and retrieved another handkerchief. In a moment he was back, wiping my eyes and trying to comfort me. "You will find that not all species are unattractive," he said, swabbing again. "One day, an alien male will please you in the same way that you and Dar find love."

"No," I said, shaking my head. "I don't want that to happen. I don't want to love anyone again. Not ever."

Shnize lifted my chin. "You do not have to love them, High One. There is no requirement that you love ANY of your husbands. You need only to …"

I jerked my head away from his hold. "Yes, I know what I'm supposed to do with them, and I don't want to talk about it. And that's an order, Shnize."

"As you wish, High One," he said, bowing formally and stepping back. He instantly withdrew into stiffness and the blank-faced, obsequious look he took on whenever he was reminded of being Franian.

"Am I free to leave now?" he asked, snubbing me with coldness.

"Shnize, stop it, please," I said, grabbing his arms, so he'd look down at me. "I'm sorry. I didn't mean to be sharp with you. Please understand. It's not that I don't want your advice. I do. I want you to talk about everything with me. I want you to be my friend. I need you. But I just can't bear to think about my 'obligations' to the RishMore. Not right now. Can't you understand that?"

Shnize unstiffened a bit. His eyes dropped to search mine, studying me as if I had the keys to the universe. "I am only a Franian, High One. I have nothing to offer you. My allegiance is to RishMore. I am honor bound to them and to you, but my soul does not contain those depths. I am not knowledgeable or competent in the elements of government, and I am not suitable to be a 'friend' of the High One."

"Why do you say that?" I asked, taking his hands in mine. "You've already proven to be my friend. Why do you worry about RishMorian codes and beliefs?"

"Why do you not?" he asked, burning my eyes with the intensity of his look.

Terriff

The wife of the High One enthralls me. I hardly took note of her before. She was an appendage to his arm. I'd remembered noting her large eyes and their odd color and that she was diminutive, weak, and easily crushed by a word from her husband. Of course, I noticed her hair. Hair intrigues us RishMorians. We should find it appalling since our females are without, but we do not. Those planets with haired females are often of primary interest.

The High One made us travel far for this wisp of a nothing at his side. Then he catered to the Terrans, dining them overly long. At the time, I wondered at that. I knew he did so for his bride, yet previously I saw no reason beyond the delicious sweep of her hair pouring down her back like free-flowing liquid and the soft fragility of the bones beneath her skin.

I had never heard that her species were ESPers. My recent meeting with the woman acquainted me fully with that fact. No wonder the High One went to such effort. The female's brain is even more delicious than the length of her hair. I fondled it, playing with its waves and convolutions. I felt the feeble rejoinders of her innocence. She delighted me.

Only then did my eyes encompass what other delicacies her young body contained. She is too small — something that has in the past been offensive to me. I have always wanted my women to stand of equal stature, with breasts pressing against my chest in the rapture of our moments. That would not be possible with this little one, yet I could lift her up. I could play with her in ways I have not played with the others. I amaze myself with my thoughts, but I like them.

During our meeting, I discussed my plans for the battle on Crezand. She listened with those eyes of hers, those large, brown and green-glass eyes. They shone as if she carried lights inside her. It drew me to inspect the rest of her face.

Her nose was more dominant than ours, but in a rather attractive and saucy manner. I wanted to touch it, to kiss its sides, to acquaint myself with the slight freckle she carries on the right side of it. That freckle maddened me. It should not be in plain sight. It should be hidden where it cannot make a male yearn.

Then she smiled. I was still getting used to her pale skin, skin without the healthy, blue tint of ours and minus the sparkles of our skin scales. Yet when my hand brushed (accidentally) against her face, I discovered that her skin was the softness of the finest material, of a baby's face when it has not yet formed the scales of maturity.

But there is nothing of the child in her. Her body peaks quite womanly. Her curves are not that different than one of the RishMorian females. Her spirit is clothed in a body made for supreme sexual gratification.

I sigh. She smells delightful, too. The fragrance of her hair and skin reminds me of home: wheat fields, green grass, fresh air. The first few times, I shut my eyes not to reveal my thoughts when I breathed in her scent. My face was steeled properly, but the Franian saw. He is wise in the ways of the RishMore. I do not trust him. Despite my care, he saw the spasms of my lust for the woman.

What must I do? She is wreathed in power. Dar saw to that. But one thing is certain. I will have her.

Previously, I plotted murder. It was not that difficult, a grieving widow suffocating with her two sons from a broken vent in the nursery. What a pity, I would say, and how easily done.

Yet, I will not kill this woman. There are other answers when a female calls to the senses, as this one does to me. I will know her, drive myself into her, and then…

I sip my drink and mull over the situation. Captain Jertot sits beside me. He has not a vial of passion in his heart. He is immune to her, I am sure. Does he not see the potential in her? How can he not? How can he not look at her and desire her?

He glances up, straight into my eyes. "You have chosen the avenue of attack?" he asks. I shake my head and glance back down. "I was debating alternatives," I tell him. Then I laugh inside. May he remain ignorant of my thoughts. Perhaps, he might be swayed by her beauty after all. He likes her; I know that. They smile into each other's eyes like conspirators. No, not like conspirators, that is incorrect. She has not a devious inch of flesh. She is as pure as the summer breeze, warm as …

What is wrong with me? I must get past this. I cannot sit here like an inexperienced youth and moan over a piece of alien bed play.

But the woman has a brain. That I well observed. The High One chose well. It surprised me, to be truthful. I knew that our former Great One was brilliant. I never would have attempted to rise to power, if he had not foolishly passed his position to an alien woman. Ah, but what a woman. I smile, despite my resolutions to push thoughts of her aside. My body aligns itself for sex, but I have no time for that, nor inclination… The one I want I cannot yet have.

Plans change. Who's to say this is not the better course? I shall rule the RishMore and take the Earthling woman as my slave. Let her lick at the trappings of her former place. Let her beg for the droppings that I give her.

She portrayed herself as haughty, but I have seen inside her. She mirrors sweetness and the naiveté of youth. She will fall when my boot pushes. I will step upon her soul and lure her into the depths of wildness, and then I shall enjoy my play.

Shnize

I escort the High One to her children, my eyes scanning the halls for trouble. I have decided to talk with Frick, but I know he will not like it. He will not listen either. I am positive he will be too stubborn.

While Prana plays with the little ones, I lead him aside and make my attempt. "We Franians must unite," I tell him as he watches the closed door, not even glancing at me as I begin my speech. "The High One is in danger. Terriff, the Minister of War plans to toss her down and turn her into his personal slave."

Frick growls. Then he frowns, and I see that his eyes grow dark with anger. "You impinge on my territory, Shnize," he hisses. "You are nothing more than a positionless Franian. What do you know of it? No one harms the High One while I stand guard. She is safe always. Do not tell me of the politics of the RishMore. That is none of our business."

Frick insults me purposefully in calling me positionless. I draw back and signal the offense, but then I reconsider. Prana's life is in danger, and without a doubt her sons will die if Terriff has his way. I need justice against Frick, but not if it risks Prana.

"I will not fight you now," I say, loudly enough for the others to hear. "We must first build a net of safety. We must cosset the High One and her children. Do you refuse to protect them?"

Frick's eyes glow with anger sparks. Since I speak in front of the others, he cannot back away. He must answer me. But he does not. Instead, he thrusts at me. I dodge. He thrusts again, but he leaves his left side open. I can easily stab him, but I only feint, showing the others that I am capable of killing but not willing.

"Who sides with our Queen?" I ask the others.

Of course, they all respond to that. But Frick feels the humiliation of losing to me. He snarls unnecessarily loud. "Our hive Queen is not at risk."

I win the others over until his words stop them. They look at me and pause.

"We form a new hive here," I say. "A hive of loyalty, a hive of honor. It is the High One who owns us at the beginning. He entrusts his wife to us. Now she is our Queen."

My words are heresy, but I must say them. I must make the Franians see.

"He spews contamination," Frick cries out. "He believes the alien woman is our Queen. She is not. She is only the wife of a dead RishMorian leader…"

"No. She is more," I tell him, and my eyes capture the others. "She …"

Prana opens the door, perhaps wondering where her guards are. "What is wrong?" she asks me, her eyes full of concern. Then she looks around at the guards, for they are staring at her, peering inside of her, studying what they have not searched for before.

It is too late for me to stop what I am starting. I do not previously consult with Prana as is wise. I can only hope that she can understand and follow my lead. She is quick. I know that.

I seek to unite the Franians behind your rule, I inform her.

"My Queen," I call her, and she turns back to look at me. "Freggggeeetth snettttt swaalll," I say.

She does not speak our language, but we are in harmony. She places her hand on the arm that I have held out. She does not understand what it means, but the others speak among themselves, angry at first, and then they turn and look at her again.

Prana's head tilts. It is a common thing for her to do. I notice it often. She does it when she is puzzled. I believe she is trying to read me. I nod. "Yes," I whisper, and I give her free access.

Her thrust inside me is as light as she is. I welcome it, smiling into her eyes. There are too many layers inside me for her to understand much, but she trusts me. I have no doubt that she will do as I ask. I know her well.

"Dresssseeellllt fretth naactth swemeeepth," I tell the others, and again their eyes dim with doubt, but they look at Prana and see the quiet of her eyes, the tilt of her head — so much like a Franian Queen's.

Frick is the only one who dares to question it. He attempts to speak our tongue, but I stop him and remind him that the High One is from a different land.

He changes languages, but he does not alter his tone. "Name the cornerstone of civilization," he demands of Prana.

I do not react, although his manner with her is insolent. I watch the others. They gasp, appalled by his behavior. If she is a Queen, his demand is cause for his death. Likewise, even if she is not our Queen, she is still the High One. One does not demand things of the great.

Prana looks at me, her tilt of head even more pronounced as she attempts to understand what is going on. I urge her back inside me. "Look," I send through our mind link. "See the words. Read them. Speak them aloud."

She nods. She studies my eyes a moment and takes a breath as if steeling herself for another problem that must be dealt with. Then she does exactly as I ask. "Obedience to the Queen," she tells the males around her.

It is perfectly done. The guards fall to the floor, not a one of them doubting then that she is our Queen.

Prana waves them up, still watching me for cues. I nod. "We protect you, my Queen. And we protect the hive sons from the warrior, Terriff."

"Go back inside," I send to her, and she nods. Without a word or even a curious look, trustingly, she re-enters the room. I watch as the door swooshes closed. Then I turn to lead the others in commencing the weaving of Prana's net of safety.

Prana

That was the strangest thing I've witnessed in weeks. All the Franians were focused on me, and Shnize was obviously their leader. I've never seen him like that before. I knew he was smart and skilled with arms, but he impressed them with whatever he said. And he was assertive and bold with them as well as with me. Is that a good thing?

I think Shnize was enlisting them to help me. I didn't really understand that, though. Aren't they already guarding me? Yet, it was like he woke them up. And they looked at me, and they saw me, *really* saw me.

Frick just entered the room. The others remained outside in the hall. Where did Shnize go?

Frick checked me over, like he always did, but on his face was a look I've never seen before. It was like he was half afraid of me or awe inspired.

I'd started to believe that Franians were like docile robots that only came alive when something bad happened. Things were never what they seemed. I've been finding that out a lot lately.

I hoped Shnize returned quickly. I didn't want to leave the nursery without him. With all the talk about Terriff, I didn't trust anyone else. No, that wasn't true. I trusted Captain Jertot. At least that made two males I could count on.

My little son, Frawn, just put his arms around me and gave me a huge kiss. He was such a love. He was given his father's eyes, the skin color of the RishMore, and the baldhead, but I saw a little something of me in him. He looked at things as I did. He studied people. That, Frawn got from me.

Little Strandor was starting to walk about. He took miniature steps and wobbled a lot. Then he plopped down and sat, looking so surprised that his legs hadn't held him for a longer time. Strandor had a happy personality. He was satisfied with less than Frawn was. My older son was driven already, like he already knew that he would one day be the High One. How could he know that? Was it really in his genes as Dar used to tell me?

Nethre, the Second Head of Nursery, brought in my Franian boys. They were the ones that Dar gave me at each of my son's births. Jason, Todd, Eric, Steve, and Alex were the older ones of Frawn's age. They were much more developed than Frawn was and appeared older than my son.

These Franian boys were standoffish with me, although I insisted, they come over and shake my hand each time they visited. I wished I could be closer to them, but Dar hadn't let me, and now, I think, it was too late to change that. They knew "their place," as Dar used to put it. How is that possible? They were only little children.

The little ones: Flan, Simon, Krall, Degat, Matt, Ben, Conner, and Thorne saw me on the floor and came running. They had no doubts that I adored them. Shnize always frowned when they did that, but he didn't correct me like Dar used to. Matt and Degat almost drowned me in sloppy kisses before Frick ordered them away.

"I like it when they kiss me," I said after Frick spoke sharply to them. I think Frick ignored my words. He rarely replied when I spoke — only when he was forced to do so. I remembered when Shnize used to be that way. Would Frick change as he got to know me better?

The boys were regarding me with soulful, sad eyes. "Come here," I called to them, and they walked closer, but they didn't throw themselves at me as they had previously.

"What did you say to them?" I demanded of Frick.

Shnize overrode my question, answering instead. "He says what is necessary, High One. The Franians must learn their place."

I stood up and walked over to Shnize. "Where did you go?"

"Leave us, Frick," Shnize ordered, and the guard, without even blinking at the slight, withdrew.

"You do well out in the hall, Prana. I want to explain."

The children were chasing each other, playing some kind of mad version of tag. Nethre was attending to them, but the noise level was quite frightful. Shnize steered me to the other side of the room.

"I claim you as our Queen, Prana. I must now inform you of our culture so you can convince the doubters."

"But I'm not your Queen," I interjected, not understanding where Shnize was going with it.

"You are my Queen and the Queen of the others henceforth. You *must* be for your own safety and for your sons. Now listen. We form a net of safety, which is the rallying of all the Franians. They become cohesive under your banner. Then, the children are safe."

Relief washed over me. I hadn't noticed I'd been carrying so much worry. I reached out and touched Shnize's arm. "Thank you. I owe you everything. What can I give to you? Name it."

He was staring at my hand on his arm. He didn't speak. Did I say something wrong? Had I hurt his feelings?

"Shnize, what is it?"

His eyes lifted to mine, then, and he said, "I want what cannot be. Other than that, there is nothing I desire."

I thought he meant that he wanted to go back to his home, and I knew he was not allowed to do that. He had told me so once. I nodded. "I'm sorry," I told him. "I would give you that, if I knew how."

My words surprised him. He searched my eyes, entered me, and as quickly withdrew.

"Forgive me, High One," he said. "That is not appropriate."

I laughed. Dar had searched my mind at will. It didn't hurt, and it didn't bother me, so long as it was meant for clarification and not for chastisement.

Shnize smiled, something he was doing with increasing frequency. "My Queen, you already give unendingly."

I wasn't sure what Shnize meant by that. I started to question.

"Mommy," Frawn, my older son said, tugging on my dress. "Will you come play with us now?"

"Your mother needs to return to her affairs," Shnize told my son.

"No Franian speaks for my mother," Frawn said, pulling himself up into his royal carriage pose, looking like a young rooster strutting. Usually that made me smile, but not this time. His words were too much out of line.

"You will not speak to Shnize like that," I ordered. "He is my Prime Minister."

"A Franian?" Frawn repeated as if the concept were impossible for him to understand.

"And why not?" I asked. "Is Shnize not special? Do you doubt his intelligence or his integrity?"

Frawn wasn't used to my speaking to him like that. He tilted his head, just like I did when puzzled, and then said, "It is not regular, Mother. That is all I meant. What would Father have said?"

Shnize interrupted just as I was about to say, "Uh" a couple of more times.

"Your father, before, often says, 'A thought can be taken in one's hand and rotated so one can view all sides. Often that provides new outlooks.' "

"He said that?" Frawn asked, his face breathless with the wonder of hearing his Father's words.

"Before the After, he often says that. I have many more of his quotes in my memory. Perhaps you would like to hear others sometime."

Frawn nodded energetically. Then, my son bowed slightly. "Yes, I would like that very much, Shnize. I would be honored by it.

"Mother," my son continued, turning back to me. "I thank you for your correction. I was wrong. I will rotate my thoughts from now on."

My older son was always stiff like that and formal. My heart ached. He not only looked like Dar; he acted like him. In fact, I was sure that my little boy must be exactly what Dar was like at age six. I sighed heavily, the sense of loss flooding me again with sadness.

Shnize waited a moment, looked at me, and then touched my elbow to indicate that it was time to go. I nodded.

"I have to leave now," I said to my sons, and I hugged and kissed them, telling them how much I loved them. (I would have done so, also for the little Franian boys I'd adopted, but Shnize had already ordered Nethre to take them away. I could tell I was in for another lecture. I shrugged. What else was new?)

We joined Frick at the door. He didn't meet our eyes and pretended that we didn't exist, except that as we passed through the door, he followed us. The other guards fell in around. As usual, as I took the first two steps down the hall, I could no longer see anything except tall Franians with burly bodies.

"One of these days I'm going to get stilts," I complained to Shnize. He didn't respond, but the rare smile came back. He flashed it once and then flitted his eyes about the hall, guarding me as if Terriff were about to attack that very moment.

We made it back to my quarters without encountering anyone. The door slid open and closed behind us, still without Shnize speaking or relaxing his guard.

"Did you really expect Terriff to launch so open an attack?" I asked.

Shnize turned to look at me. "I always expect an attack, open or covert. It makes no difference to me. I am always ready."

I knew that, but it wasn't what I meant.

"You must eat now," he said, gruffly, changing the subject.

I was the ruler of three galaxies, and, yet, my servant ordered me around. I suppose I should have asserted myself, but I didn't. I suddenly dropped down on the bed with a bad case of giggles.

Shnize had seen me like that before. He gave me an odd look as if wondering what had set me off this time, but he didn't stop to ponder it. He moved over to the food machine and ordered our meals.

"Make mine a hamburger and fries," I called out, amid bubbles of laughter just short of hysterics.

Shnize paused and searched my face for meaning, but when he saw I was still under the influence of a steady stream of giggles, he proceeded to order drinks and the usual vitamin enriched fruit and vegetable *Blooka*.

Terriff

There are many who grumbled about the choice of the new High One. I first heard their complaints the day after Dar's death. The crew's faces were wreathed in shock then, and no one was thinking properly about whether their dissatisfaction was appropriate (or even wise) to voice. A lot of things were said in those hours that wouldn't have been whispered, normally. I knew that, yet the eruption of emotion set my thoughts churning.

From such uninhibited honesty, new bridges are spanned and new ideas are fostered. Into the medium of that unrest, I, at first, only dropped hints. I was to be the new Minister of War. What should I do if the new High One didn't wish to fight the Crezand? I asked.

Of course, I knew what they would say. It fell together like matching pieces. They pushed me into it. I just sat back and said a humble word here and there to keep it going. But with my willing ear, the objections continued, mumbled, but clearer each time and more definite in meaning.

"That woman is an alien," I heard repeatedly and nodded each time. "She is unschooled in the ways of the RishMore," others told me. "She is no older than a child," declared the ones who remembered that Dar had ruled for more than five hundred years.

There were several voices that complained that no woman could hold the RishMorian kingdom together. Then there were my favorite doubters, the ones who proclaimed, "She was only a 'play toy' of our besotted former ruler; she is not a fit leader for the RishMore.

Each time, I sat back and said, "Is that so? Really?" And in their eagerness to convince me, they persuaded themselves. Manipulation is so easy for an ESPer of my degree. So easy, in fact, I could scarcely keep from laughing.

I agreed with all the statements, making sure that I gave each speaker credit for being so astute, and thus, I encouraged the raging conflagration to spread. Sometimes, I even blew quick, short bursts here and there when I saw a danger of the blaze dying out.

Thus, I multiplied, strengthened, and spread the bits and pieces until the kindling roared into a raging fire. Then, through carefully staged meetings, I aligned key players to my side and watched as the inferno grew in strength.

Poor little alien, I'd thought, yet I'd formerly had little sympathy. I had not seen her then. I had only heard of her long, gold streaked, dark hair and the way her skin supposedly looked like the first fall of snow yet was warm and soft to the touch. I had heard that story, but

I'd never had a chance or the desire to personally check it out. What was another wife of the High One to me?

Then when I met her, I saw. The stories didn't half tell the truth of that woman's beauty. She was the hunger before a big meal, the thirst before the first sip of liquid… She was everything a man could desire bundled around a quick mind and nicely tempered sweetness.

My thoughts were suddenly jarred as I heard "Terriff." One of the co-conspirators was yelling for me. I sighed and pushed away my thoughts of Prana.

Henrick strode forward, his gawky elbow knocking off a paper from one of the tables. The owner looked up with an expression of irritation, but seeing Henrick's massive size and recognizing the status of the uniform, he simmered down and retrieved the errant paper.

Henrick, not even aware that his gangling and careless manner had pushed against several chairs, stepped on someone's foot, and caused the robot to drop a meal, continued towards me. Eyes followed him, but not a single person complained. Henrick was that kind of warrior, the kind with whom it was usually better to avoid eye contact.

At last, he arrived at the table where I was sitting. I shook my head warningly and made a gesture of silence since he was once more calling out my name. I had chosen to meet in the *Flerw* because it was one of the small, secluded dives on the bottom floor of the ship that few frequented — chiefly because its food was only passably eatable, however, without the dining hall, many of the elite had begun frequenting smaller food depositories. It was my hope that none of them would drop into the *Flerw,* but one never knew. Fate was quirky.

It was interesting to me that places such as the *Flerw* were always located deep in the bowels of a ship. It was certain that they served a purpose. Every crew needed somewhere for dark and secretive matters to be discussed. After all, where else could the seeds of treason germinate so readily?

“Well, did you meet her?” Henrick demanded impatiently. “Did you see the alien who thinks she can rule the RishMore?”

I didn’t say anything as a server was just placing my drink on the table. I probably could have kept talking. Servers are just robots, but I liked to be cautious. Treasonous talk can actually get a person killed.

The robot rolled away, and Henrick raised his voice even louder. I said, "Did you get to visit with Dar’s little play toy?”

I hushed the man again and glanced all about me. Nobody was looking at us. They were all involved in their own conversations, but I couldn’t be too sure. The High One had found ways of listening in on the crew. Would my sweet, lovely, little alien observe that tendency? She was an engaging little scrap of femininity, it was true, but inside that delicious case of hers, I’d just learned that she had a keen brain.

“Let’s call her ‘the trinket,’ shall we? That way, in case anyone should overhear us, they would not connect our conversation with well — that one.”

“All right,” Henrick said, wagging his ears a bit, as if my prevarication had upset him. “She’s the ‘trinket’ from now on. No problem. Now give. What was she like?”

I stared into the saffron-colored eyes of the Chief of Armaments and wished once more that I didn’t have to deal with him. He made my skin itch.

“She is different than I suspected,” I hedged. “More astute, more capable.”

“You jest. She is only an alien, nothing more than a hostage wife.”

I nodded. I could not afford to alienate this man. But his words about Prana did not ring true. Why had the Great One chosen her? Why had he made her First? I wanted to know more about the woman. I yearned to know more — intimately.

That skin of hers, so dissimilar than my sparkling scales, was it really as soft as they said? What would it feel like to hold her in my arms? To feel that smoothness against my naked skin? Would she bruise easily? Would my scales scratch her?

Her mouth was rounder than ours, yet not at all displeasingly so. I saw the pale pink of her smallish tongue when she wetted her lips. The memory of that hardened me. I wanted her.

"Are you listening?" Henrick demanded.

I looked up. I had received too much training to give away my thoughts. Henrick would never know them by reading my face, but he was peering into my eyes suspiciously, as if he could.

I took a drink. "I was only pondering how best to go forward with our plans," I lied.

The darkening of the man's eyes gave way to light as he considered what I'd just said. "Ah, yes. That is okay then. I thought you might have reconsidered. It entered my mind that, for some reason you might have been taken in by her exotic beauty."

"Have you ever seen the High… the trinket?" I asked.

Henrick laughed a high-pitched cackle. His overly large mouthful of blue-greenish teeth appalled me. I looked away.

"No. Does it matter? What should I care what she looks like? You see one alien, you've seen them all. Weird. Strange. I spit on them."

I nodded. "I see." I nodded again as if in agreement, then said, "I agree normally. But alien or not, this one is beautiful, and beauty draws sympathy from people. I have had to reconsider our plans. It will be necessary to alter them somewhat."

Henrick had downed his jug of brew a few minutes past. He signaled for another and glared at me while he was waiting. "What kind of alterations do you mean?"

The robot slid back to us. I paused, waiting for it to leave.

"Bring me a plate of douvmesh and jeoprrhg," Henrick ordered. "What do you want?" he asked me.

His order sounded good, but the memory of little Prana swam across my vision. If I ate too many plates of douvmesh and jeoprrhg I'd crush her. I sighed. "Bring me snodbry and vegetables."

Henrick gave me a look. "You dieting?"

I patted my tight belly. "RishMorian leaders are never fat, remember?"

Henrick threw his head back and laughed full-throated. "Better you than me. I need my fat. It helps me dominate."

In my opinion, the man needed a little less domination. His pouch already sagged outward and downward, but, of course, I said nothing. I shook my head when he asked me if I wanted a second drink and thought about what that pale, white skin of Prana's would feel like underneath me.

The meal passed with little accomplished. I explained the new changes in our plan, telling Henrick that I did not want to push forward quite yet; I needed time to gather more backing. Besides, I wanted to see Prana again, but I didn't tell him that part. Could it be that I'd imagined that heady scent of hers? Was she really as desirable as I recalled?

In between my musings, I encouraged Henrick's preparations for the upcoming battles on Crezand, giving him input from my discussion with the High One. Crezand would be the place I would make my stand, I told him. A war hero would immediately gain the people's praise. My victory would establish me clearly in their minds. I needed Henrick for that endeavor. He would be helping me with the weapons. So, I would endure his bad jokes and his nauseating table manners and continue to stroke his ego. But all the while, I would continue to dream of the lovely Prana.

Shnize

My mind is a kettle brimming with plans and plots. I know that I move at the right time to mold Prana's Franian guards into a combat-ready force with the net of safety securely around our Queen. My fellow Franians are, even now, linking every Franian on the ship into a tight Queen's guard.

Every Franian, whether he is in Prana's personal force or not, will soon vow to put her protection as his foremost purpose. I feel no guilt over my methodology. Prana *must* be our recognized Queen. She deserves it.

I glance at her face as she lies giggling on the bed. Her brow and the area around her eyes hold signs of fatigue, but she still glows from the hugs and kisses of her boys. I can tell her thoughts are of them. She has the look of love a woman gets when she thinks about her children.

Since my Queen is well occupied, I allow my mind to wonder. The guard, Frick, concerns me. His posture on our return to the Queen's Chamber indicates contempt. I need to work some more on him. I must bring him into "Queen adoration." I must be able to count on him, or if not, I will need to place another in his position.

Minutes before, the corridors we pass through are empty. It is as we find it so often these days. The echo of our feet resounds sharply against the silence of the halls. Each time I escort my Queen, I project outward, listening for attackers in wait. Terriff does not yet begin his traitorous plans. That is good. I require more time to develop the net and to get the Franians into position.

I glance over at Prana again. Her giggles are dying down as fatigue takes over. I carry her the dinner and beg her to eat. She sits up with a wilted look, sighing.

"Don't you ever laugh, Shnize? Don't you ever feel really good and want to have fun?" she asks me.

I set her plate down on the table that I've lowered into position. Then I return to get my serving, remembering that she will not eat unless I do.

As I sit down in the chair opposite her, I drag through the days of my youth in order to please her with memories. I try to remember an occasion, but I cannot. Even our play in childhood is sober. Laughter is not common for Franians, as it apparently is for Terrans.

"As a child and now, I am content," I tell her. "I take pleasure in serving you well."

"Yes, but what would make you the happiest? What would make you want to laugh from the pure joy of it?" she asks, tilting her head in a Queen's manner.

I notice how she uses her fingers to eat the croodles. Her long, delicate fingers lift up each strand and lower the end down into her mouth. Then she nibbles, lowering each piece downward. She eats like a child, playing and savoring the taste of her food. I envy that ability of hers to take pleasure in such things.

I cut my croodles and take them in large bites. They bring my taste buds scant pleasure. I suppose croodles are tasty, but Franians eat for nutritional purposes, not for enjoyment such as Prana finds.

I recall her query. She always asks me difficult questions, probing, trying to understand me. I do not comprehend such things. How can I explain to her?

"Franians do not feel the way you Terrans do. We serve. We know honor. We take pleasure in being good servants."

"No! Forget all that," she says. "You're capable of smiling and finding humor. I've seen your flashes of amusement. You just control it. You don't want people to see."

She is correct that she often amuses me. Before the After, she amuses Dar as well. He does not laugh before Prana arrives. Yet, the realization that I am also changing because of her presence is a bit disconcerting. I stop eating for a moment to analyze that.

She swings her legs up on the bed, shifting her body in a way that draws my eyes. Then she rises to stand, staring down at me. "Let me try an experiment, ok?" she asks.

I nod automatically. Of course, I do whatever she wants. She walks around the table, and then approaches me from behind.

"Do you wish me to sit or to stand?" I ask.

"Stay put. I'm going to see if you're ticklish."

Then before I can react, to pull away from her, to scold her for touching me, she places her hands at my sides and begins to jab at my ribs.

I catch her fingers and stop them. "I am not ticklish," I say. "And it is not appropriate to touch a Franian in this manner."

"Why?" she laughs, her eyes flashing the same strange lights as I have seen in them when she plays with her two young sons. Play. She is attempting to play with me. Thusly, I understand her purpose. I return her hands and ask, "May we resume eating now?"

I see into her eyes at that moment and witness the light go out. Sometimes, when we are planet side, the sun shines warm and bright, and then a cloud passes and momentarily absorbs the light. So, it is with Prana's eyes. My refusal to play brings a cloud across her eyes. Yet, I cannot alter what I know to be right. If I allow her to touch me, I fear I cannot control my body's impulses.

Turning away from her disappointment, I take another bite. I am not ready for her second attack. She catches me by surprise.

"Auuuh," I cry out the moment I swallow. Her small, agile fingers are once again probing my ribs in a manner most irritating and yet most pleasurable. "Auuuuh," I howl and stop her tickling fingers.

She cries out then. I have been too rough.

"Forgive me," I say, automatically bringing her fingers to my lips for the comforting kiss I have given her boys when they hurt themselves.

Her eyes follow, watching. She wears an even stranger light in her eyes. "You laughed," she says. "I saw you. You actually giggled. You *are* ticklish. Let me go. I want to do it again."

It is obvious that I do no harm to her fingers. They are straining against my hold.

"Prana, stop this. It is not appropriate for…"

But she does not listen. She jerks her fingers away and attacks me once more. I have no choice. I rise up, secure her arms at her side, and pick her up. Then, I walk over to the bed and drop her.

"I must leave you. I cannot do this," I tell Prana, more sternly than I feel. Her face falls into gloom. The words cause her pain. I see that, but I cannot play this game with her. It is not in me. And even worse, her childish actions make my blood sing and my scepter throb. I bow to her and walk away.

Prana

My silliness drove Shnize away. I tried to stop, but I was so lonely. "No, please, wait," I cried out, dropping my arms to my side. "I will behave suitably. I will sit down and eat my dinner… if, only you will stay. Please?"

I could see that he didn't want to. His eyes were staring into mine, and for a moment I thought I saw emotion in them, something almost like passion, but, of course, I was mistaken. He was Franian, a fact of which I kept reminding myself. I sighed and behaved as I should have. The bed sank down slightly with my weight, but then firmed up to accommodate my sitting position. I reached out and picked up my utensil, took a bite of the plenk and flrerst, and sighed again because Shnize had once more sat down.

We ate in silence for a while, I because I was afraid to say something that might send him off, and he, because undoubtedly, I had disgusted him, was putting up with my presence simply because he wished to serve.

I couldn't help sighing several more times before the meal ended. Finally, he spoke. "You are unhappy. I am sorry, Prana. Are you dwelling on your loss?"

To say I didn't love Dar would be a lie. A woman always feels love for the man who first beds her, fathers her children, and becomes her companion. Yes, I had loved Dar, and I missed him, but as the days went by, I found the love part less than I'd imagined. Dar had been a whirlwind who swept me up inside his gyrations. He had carried me into new worlds; he had consumed me, but I was surprised how infrequently my thoughts drifted to him.

Shouldn't the absence of love be like an amputated limb? Shouldn't pain be constant, a loss so numbing that all thoughts of everything were only background noise? Yet, it wasn't like that for me. Maybe I was incapable of love, having had so little of it in my former life. Maybe I was damaged inside.

Yet, I couldn't be honest with Shnize. I couldn't admit those thoughts. I just nodded and stayed silent. It was best, I guess, to hide such self-doubts.

The rest of the evening, Shnize was gentler with me, as if he took my silent nod as the witness of a heavy heart. It made me feel worse for deceiving him, but I could not tell him the truth. I just couldn't. I was too ashamed of it. Besides, I really had loved Dar, hadn't I?

Chapter Six

Frick

I stand in front of the door to her chamber. It is my job to guard her, but things keep shifting. I feel unbalanced by the tilt of the floor of my beliefs. In the past, our hive Queen tells us that change is like a fresh breeze. It airs the hive and brings new life. "Change," she says to my childhood self, "is the uncurling of the present into the future."

I hear once that a hive Queen can read the future by the breath of the wind. If so, does she know about this High One? Is it possible she anticipates the small alien who Shnize now calls our Queen?

I do not understand the changes on this windless ship. I am sure that things are not the way they should be. A master does not blow up and depart for the After. A lowly wife does not become a Queen. A Franian does not gain a position at her side. These are wrongs. I feel them, but I do not know what I must do. She passes the test. She is a Queen, but is she our Franian Queen? Do I owe her my allegiance?

There are too many currents that confuse me. They flow past me in waves and dashes. Why does the High One, this woman child of the late master, follow the commands of a Franian? I see her listen to his guidance. I see this with my own eyes. She obeys him as if his commands are those of her husband's. In fact, I believe she bends to Shnize's wishes even more than she does to Dar before the After. Why is this? What brings these departures from the norm?

Shnize acts very unFranian, too. It is nothing that I can drill a hole with, yet I sense it. He is changing.

I recall how he took the assertive with me, ignoring my authority. He, knowing fully well that I am the High One's Sentinel now, demands MY obedience. That is against tradition, yet he breaks the sequence of command. The others do not interfere. Why do *they* allow it? Why do they not demand the "right of position?"

And then, as if the winds of change do not blow down all protocol, Shnize fights me. It is rare for Franians to clash outside our training arena. We do not. We compete with others only to improve our skills, and to enhance our position, but only in the privacy of our practice arena.

But he challenges me during our worktime. and he lashes out with full vehemence. He does not fight like a Franian. He battles like a RishMorian. That is another difference.

Yet, now because I lose, it obliges me to follow him. Shnize wins my allegiance, and I cannot argue the situation, yet it all feels unbalanced and wrong. Now he calls the High One our Queen. I do not believe she is, yet I cannot question him further, nor can I stand against him — or the Queen.

I am stuck in the unevenness of variation. I must accept, but I feel great bewilderment. An ocean rolls with turbulence, but our lives have an obligation to remain placid. Duty centers us with its constancy.

Prana

The days were passing, but there were no leaves on the trees to watch turn green and then gold and brown. There weren't even any trees, except the artificial ones in what Shnize called the illusion room. There were no changes in the weather or in the length of the days. There weren't even any holidays to give the passage of time that special spice of anticipation.

I planned to instigate holidays when I was on firmer ground with the citizens. But Shnize said that I must not make too many changes. He told me that moving slowly was safer. It was hard to move slowly when I was still young. I wanted to live life more fully than just studying war plans and discussing governments.

We met with Terriff again yesterday. He was wearing a most exquisite deep burgundy color in a smooth-velvet-like coat that draped down over his buttocks. I wanted to pet the fabric the whole time he kept talking about which city to bomb first. I imagined the coat would feel like the nose of the toy animal kitty I used to touch sometimes at the institute. But Shnize was glaring at me just for thinking about it.

Although Shnize said he wasn't able to read my thoughts, I think he could. I remembered how my husband could even finish my sentences sometimes. He saw so clearly into my mind when he wanted to. Perhaps, Shnize couldn't do it as skillfully as Dar. Maybe. Shnize said it was only my expressions and postures that informed him of my thoughts, but I guarded them now. I think I'd gotten better.

I was even more careful of Terriff at our second meeting. I felt the strength of the power he used to bat against me. It seemed much stronger than before. That's when I discovered that he could read my mind like Dar used to. I made sure that I was sheltered from him at our next meeting. I took a pill that Zoff gave me. It ran interference.

Because of that, Terriff was not as friendly. He snapped at me right off for bringing Shnize. I had to retaliate with nastiness. I've never wanted to be unkind to people. That's something that made me even more unsuitable for being the leader of the RishMore, but I've been careful not to let anyone know that.

"The Franian has no right to sit in a war conference, no matter what his title," Terriff bellowed with his flat nose flaring, his teeth snarling, and his eyes flashing dark warnings of danger.

I froze when he said that, not even sitting as I'd planned. "You dare to insult the High One?" I lashed out at him without thinking. "Guards, take away this insurgent and teach him manners," I ordered.

The guards, all Franians, of course, moved forward without question.

Terriff melted down then, like a too hot waxy candle. And believe me, he cooled down quickly. In seconds, even before the guards touched him, he was pleading for forgiveness. His nose still flared, but his teeth no longer snarled, and his eyes held little dark. His whole body yellowed slightly, which clashed with the burgundy of his coat.

I let Terriff plead for a minute or so, then I looked deep into his eyes, allowing him to see that he could not read me even at the surface. When he discovered that fact, his eyes grew large, his face yellowed a bit more, and his body shrank at least an inch or two.

"Forgive me," he said, bowing his head with less stiffness of pride than he had shown before. "I misunderstood, High One. I thought you merely did not understand that a guard does not belong at your side during talks."

Frick, without glancing my way, laid his hands on the War Lord and started to arrest him. The other guards behind him moved closer, forcing the Terriff to take a step backward. I was just about to call them off when Shnize did. "Hold," he ordered. "High One, do you wish the War Lord locked up?"

As, I said, I didn't like being mean to people, but Shnize had told me such horrible things about this particular RishMorian. It was hard to back down. But Shnize was telling me to do so. I shook my head. "I will accept his apology THIS time," I said.

Well done, Shnize whispered into my mind.

Frick released his hold and backed away, as did the other guards.

Captain Jertot, his eyes crinkling with suppressed humor, sat back down, poured himself a glass of juice, and spread his legs out in front of him. “That was quite a show, Terriff. I would leave the Franian alone if I were you. The High One has made Shnize the Prime Minister, remember?” Jertot said.

Terriff shook himself once, wiped down his burgundy coat, and deposited himself in the same place where he’d previously sat. His smile didn’t look half as cocky anymore, and he seemed to have let out some of the starch from his upper body. I didn’t laugh, but I felt a little like doing so. I felt absolutely no remorse at embarrassing the man. And, if Terriff really had intentions of hurting my babies, I would get even tougher with him.

So, the meeting began with an entirely different tone than the one Terriff had expected. Shnize sat between us, a positioning I insisted on. Terriff’s eye twitched at the affront, but he said nothing. Perhaps he was learning.

Unfortunately, Terriff got over his humiliation after only a couple of minutes. With a word from Captain Jertot, he launched into a fully illustrated and up to date (according to our spies) sketch of the Crezand’s launching pad, army headquarters, and governmental centers. Then he demonstrated the battle plans with a far too vivid and concise computer simulation.

The bomb detonation covered miles of land. I watched as the faces of the fly people looked upward, realized what was happening, and witnessed their punishment. Ice cascaded down, a globe sealed up the site, and the standing people froze in various positions and poses.

We saw it occurring simultaneously in seventeen locations. Doing them all at once like that would leave the Crezands no time to launch a counterattack or to flee to safer sites.

So, time after time, the three of us observed the simulation horror from different angles. And at each occurrence, as it concluded, and the people in the site froze, an analysis spilled out, describing the various projections, accompanied by other assorted data.

Terriff's plans appeared successful, but I hated watching the effects. I knew I had to retaliate against the Crezands. It was a fact that both Captain Jertot and Shnize had drilled into me. (I could almost hear Dar's voice in my mind reminding me: *You must banish mercy when necessity demands it.)*

As we studied the final data, with each analysis and sketch, Captain Jertot gave me his opinion of the effects, as did Shnize and the by then, only slightly humbled Terriff. At times, Shnize delivered both spoken input and that which was said directly into my brain. Thankfully though, he didn't give me a lot of the latter, for I wasn't good at masking my expressions, and I definitely couldn't listen to two conversations at once.

Eventually, the four of us agreed on which sites were most effective, choosing among a selection of twenty-five sites theorized to do the most damage. Then we decided the when and how of it. I hated that part because it meant I was more or less ordering the half-death of thousands.

Truthfully, when I thought about it, there wasn't anything I did like about being a ruler. Outside of the war room's horror, I certainly didn't get any more freedom. So far, I hadn't done a single thing I wanted to do. Even spending time with my sons was briefer and sparser than in my days with Dar.

"This has brought tears to your eyes, High One. It was obviously too much for a beautiful woman to witness," Terriff said, interrupting my silent moaning and groaning. I looked up at him. Tears?

"No, it is not tears," Terriff corrected himself. "It is merely the sparkle in your eyes. Forgive me, their beauty blinded me."

"Crembort," Shnize said into my mind.

I glanced over at him. Shnize was glaring at Terriff, his eyes hurling arrows of rage.

"What does that mean?" I asked, so puzzled, I mixed up a mind-shared thought with a verbal one.

"You cannot understand my meaning? You cannot see the beating of my heart?" Terriff continued.

I darted a look at the War Lord. Cannot see what? I wondered, but my mind was still trying to figure out Shnize's "Crembort." Besides, I didn't really want to see Terriff's beating heart. The idea nauseated me.

"I have fall to you," Terriff said, speaking for the first time in my native tongue.

"Dorbemz," Shnize mind spoke, which spun me back to stare at him.

"My heart drum-thumps," Terriff said, seizing my hand to plant a kiss on my palm.

I didn't get to ask about "Dorbemz." Shnize launched himself at Terriff faster than a zimbur flutters its wings.

"You do not touch the High One," Shnize roared.

Captain Jertot rose. "Have we not done this before? Is it not logical, War Lord Terriff that you must keep your hands off the High One, so the Prime Minister does not throw you into the daggers of the waiting guards?"

Everyone froze, just like the people in the simulations.

"Shnize, we need the War Lord," Jertot said. "Terriff, will you kindly remember the position of the High One?"

The captain was speaking so rationally, it calmed down the growls of Shnize. My former guard looked at me, took a step back, then bowed slightly and waited for my command.

"He did not harm me, Shnize. I will let it pass *this time*. What is Dorbemz?"

Be silent. That is not a word you should know, Shnize whispered into my mind.

The others were staring at me. I knew I had to say something, but I was tired, tired of war, tired of doing what I didn't want to do, tired of Terriff and his beating heart.

I stood up. "This meeting is now over. It will reconvene tomorrow at the same time," I commanded.

Nobody corrected me. No one said, "You can't do that." My command was, for the first time, a complete success. The others stood back, allowing me to walk in front, that is Captain Jertot and Terriff did. Shnize, of course, walked beside me, still glaring and growling at Terriff.

Then, my perfect moment was blasted away by a computerized voice coming through the wall. "Crezand ships approaching. Analysis: warships, attack ready, carrying the following charges…"

We all froze. War. Now? The ship's voice stopped, battle analysis completed. The four of us surged back toward the table.

"Down screen," ordered the captain. "Missiles ready."

Captain Jertot had resumed his seat, familiar with such procedures from the past. Terriff, a little slower to move, was staring at me like he thought I'd run in the opposite direction.

I gripped Shnize's arm, as if I could pull courage from his strength, and in a daze, I returned to my seat. My blood sped at dizzying speeds. My heart beat so loudly, it deafened me as I stared at the three enemy ships heading for us like sharks in the water.

"Your orders?" the captain said, with a voice so grave, it sent shivers up and down my backbone.

"Report, Terriff," I demanded. I knew the sequence. I'd seen countless vids of all this before at Dar's insistence. I'd learned the protocol, just in case. I hated his *just in case.* Yet he'd drilled me with it for hours. Had Dar somehow known how essential such knowledge would be? Had he seen into the future?

Terriff shot out statistics and choices. I nodded. *Thank you, Dar*, I whispered softly. *Thank you for ignoring my stubbornness.* "Destroy this insurgence," I said. "Send them twenty light years away. No survivors."

The captain didn't pause. He must have been expecting exactly that particular command. His fingers programmed the sequence. Terriff signaled his sanction, and my finger pushed the final button. Thirteen annihilators shot out of our stern and beamed on the exterior metal of the ships and the fly people inside. Thirteen bursts of energy. Thirteen explosions, but not a sound. Our ship rocked from the waves of the destructive force, but no injuries were called, no ship's damage.

When the light show was over, the space before us was dark again. No fly eyes, no ships, no sight of any part of their existence. Just emptiness.

Reports were coming in. The most important: Survivors present, 0.

"Well done," I said because those were the words I was supposed to say. I nodded to Terriff and to the captain and stood up, my hands shaking, but still firmly gripping Shnize's arm. "We will discuss this tomorrow," I said more confidently than I felt. Then, with my head erect, my eyes on the door, I headed for the exit.

"But, High One," Terriff began.

I ignored him, not even pausing. The door slid open, and I walked through it. Then it closed behind me, and I kept going, heading for my room. But the voices of all those aliens were screaming in my head. Fly eyes like Marta had been — Marta's eyes, so big, round, and full of pain.

I stopped suddenly. Shnize wasn't expecting it. I felt his body still striding forward, the heavy muscles in his arms still rippling with the tension of my touch. I dropped my hold on him and sank to the ground. Pain surged through my head and stomach. "I'm going to be sick," I said.

Shnize was there instantly. "No, you cannot. This is too close."

I cried out when he lifted me up. The surprise of it kept me from erupting at that moment. Yet, I could feel the sickness inside me needing an outlet. "Shnize, I can't…"

He broke into a run, carrying me forward. I cried into his chest, the lump in my throat still threatening. He almost made it back to my room, but I couldn't hold it that long. My stomach burst out the foul contents in a geyser of discard. If I'd been Shnize, I would have tossed me down and thrown me into the pool of stink I'd just erupted, but he didn't. He held me, one hand stroking my head, the other holding my hair back out of my face.

When I was empty, Shnize picked me up and carried me forward, still caressing me, still reassuring me with gentle words, those that a loving parent speaks to a sick child.

I was too embarrassed to say anything. I clung, my arms around his neck and my head against his chest, limp as a pile of dirty clothing.

The door of my chamber slid open and then shut as Shnize carried me inside. "I'm sorry," I cried out, humiliated that I hadn't made it into my room without throwing up all over him.

"It does not matter, Prana. You do what is necessary. Dar in the After is proud."

I sighed. "All those people. I sent them into oblivion. They will never return to their homes. I wasn't thinking about that. Not then. I might as well have killed them."

Shnize set me on my feet. His hands raised my arms and then lifted up my dress, pulling it over my head. “I can’t…” I began, but Shnize hushed me as if I were still that child he’d crooned at in the halls.

“I am Franian, Prana. It makes no difference. Now, take off your underwear and slip into the pool. You need to wash yourself.”

Like a zombie, I obeyed him. The water was warm. I slid into it and found comfort. I floated, my eyes closed, my hair spreading out all about me. For several minutes I continued like that, and then I heard a splash on the other side of the pool. I opened my eyes and saw Shnize wading toward me.

“What are you doing? Leave me,” I cried out, suddenly coming out of my fog.

“You are tired. I wash your hair.”

I had backed to the other side. There was nowhere else to retreat, and I was naked and defenseless. Shnize was almost double my size; I knew there was no way that I could fight him off.

“I don’t want you in here,” I said again. “Get out.”

He stopped, watching my eyes. “You would fall asleep in the water, Prana. I cannot let you drown. I am Franian. Do not fear my presence. I only wish to serve you.”

He was always telling me that. Franian, Franian. What did that have to do with anything? He was a man. Wasn’t that more significant than being Franian?

My musings had given him the pause he was looking for. He waded forward and reached out for me. In that moment, I felt the coldness of the shampoo on my scalp. Shnize’s hands began to massage the liquid in. I stiffened, but I didn’t speak.

“I remove only my outer clothing, Prana. Please, relax. I do not hurt you, not ever.”

Did he know that my heart was racing, that I was breathing in short, shallow, and very shaky gulps of air?

"I am proud of you today. You handle the emergency well. Terriff wants to conquer you, but you do not allow it. He respects you more now."

"What does that mean? He'd allow me to speak sometimes when he was bedding me?"

Shnize laughed. "Rinse," he said. When I came back up, he poured more shampoo and began again. "Your hair took in some of your disgust," he said.

It was a moment before I understood what Shnize was saying. "Oh, yuck," I sighed. "It must smell dreadful."

"Not anymore," he told me, still massaging the shampoo into my locks. "It smells more like you now."

"I made a mess in the ship's corridors. What should I do about that?"

"The robots deal with it. They clean constantly. Their inner alarms tell them there is a spill. Rinse."

This time when I dipped my head, Shnize was satisfied. "Lie on your back," he ordered. "I cannot get the shampoo out like this."

"I can finish it now," I told him, not wishing to expose my breasts.

Shnize laughed and flipped me over, his hands still rubbing and stroking my hair. I came up coughing, having swallowed some water.

"Sorry," he said. "One more dose of medicine, and you can get out."

I finished choking and opened my eyes to look at him. "Dose?" I asked, croaking like the frogs outside the window at the Institute.

Shnize poured on more cold, and I understood. The tamer rinse, something my long hair had to have, or the tangles made combing impossible.

Once more, Shnize's fingers did their magic, and I waited for the command to rinse.

"Come, let's move away from the soapy water," he said. "I want your hair to rinse good and shiny."

He didn't wait for my compliance. With his arm around my waist, he dragged me away from the side of the pool into the deeper water of the middle. He stopped where he could still touch, but I couldn't.

"Shnize," I complained.

"Rinse."

The rinse was only a quick drop, swish, and up because I liked to leave some of the tamer rinse in my hair. Apparently, Shnize knew that. As soon as I was finished, he towed me over to the other side, next to the steps.

"I get out first," he said. "Then I retrieve your towel, so your body does not develop a chill."

I was shivering all ready, but I didn't mention that. However, the idea of his getting out of the water was a comforting one. I nodded.

Shnize was as good as his word. In a moment he was standing there with a warm, inviting towel, but I was still in the water, naked. "Just put it there and leave," I said.

He shook his head. "I do not trust you in your weak condition. You need food and rest. Come, I take care of you."

I could have argued, but I was suddenly so limp with fatigue, the thought of that warm, dry towel was a carrot to my need. I rose up the steps and lurched into the towel.

"Good girl," Shnize said, and he lifted me up, despite my protests, and carried me to the bed. Underwear and a clean dress were waiting for me. I stood where he'd set me and stared down at my things. Relief flooded through me, and yet, somewhere, also a stab of disappointment knocked against me. I slid my panties on, my upper garment, and then tugged on the dress.

Meanwhile, Shnize, with his back to me, was working the food machine. In a moment, he returned with a meal for me.

If anyone had asked me about hunger, I would have laughed in his face. I was still nauseated, and my stomach hurt, yet the sight of the food that Shnize had chosen, sparked my interest. It was a bland meal, pureed fruit, and noodles in a non-spicy sauce. I ate a bite and then some more.

"Thank you, Shnize. You were right. I do feel better."

He gave me one of those flashes of smile and nodded. "You do not order me to sit in a *dungeon* because I do not heed your directives?" he asked.

He'd used the Terran word "dungeon," apparently having nothing equivalent in RishMorian. I laughed. "Where did you learn about dungeons?" I asked, but I continued to eat, trying not to remember the feel of his hands massaging my hair or his arms around my waist as he'd moved my naked body into deeper water.

Terriff

She handled herself well. I was disappointed. I'd envisioned her disgrace and her inability to function under pressure. I'd thought, also that she would be too weak to carry through, but she did not react predictably. Instead, she was a RishMorian in every way.

Days before, I had discussed with Jertot the new High One's lack of qualifications to lead us into war. I had been interested in noting his reluctance to state his opinion. I knew he liked the little alien. I knew, also, that he respected Dar's opinion. Yet, I'd thought that his vote of confidence was only blindness. I'd even believed that Captain Jertot was prejudiced by the woman's beauty. I was wrong in both accounts. The High One had spoken exactly as Dar would have. She could have been his clone.

Did that leave me without a plan? Of course not. Worthy or not, she was still an alien. I deserved the position. I would bring fame to RishMore. I would give them new victories and a larger destiny.

I sighed. What a shame. Such a woman. To have her at my side, cheering me on, warming my nights, feeding me her youth with her sweetness and trust. Ah, yes, I understood Dar well. I would have chosen her, also. She was a woman worthy of love.

Then, a novel idea flitted across my brain. What if… I nodded my head, scratched my chin, and played with the thought. I closed my eyes and dreamed it bigger. She was still young enough to spawn more. She was healthy, and she represented a planet not yet suitably subdued. Maybe it would work. But her sons. What would I do about her sons, then?

I stood up and went to look out the panel window. The ships of the aliens had left no trace. Their war ships were far off in a distant galaxy. No doubt their captains were wondering where they were, debating what they should do. What a shame for their loss. But life was like that. Separation. Closure. Parting from the past.

The passing of Dar's boys would be equally quick. The tweak of an eye, and they would be gone. Then how grateful Prana would be when I gave her another little one to love and hold. Yes, there was merit in my idea.

But then another obstacle passed before my eyes. Shnize. The Franian irritated me. I must deal with him. But how?

I could see that to separate her from the Franian's side would spin her into weakness. Then she would fall like an unpropped board. Yes, his death must come before all other actions. The Franian would be dealt with first, next the boys, and then I would weave the woman deliciously into my life, perhaps even grant her a title, if she deserved it, if she served me well.

With a smile, I left the War Room and went in search of the members of my mutiny.

Frick

I hear the tale of what she does while I eat my meal. The rumors are flying. She scores big against the Crezand. Everywhere I listen, her name is on the people's lips. She holds their admiration as if she engages in hand-to-hand combat with the disgusting creatures.

I think about their words as I spoon my cbrun stew. Can it be that she is as worthy as Shnize tells us? Is she Queen material? Can she be the foreign Queen from our legends?

My thoughts are on her as much as on the food I eat. I ruminate and then look around, co-mingling old thoughts with new, clearing out foggy patterns, imprinting them with possibilities.

Shnize

I leave her now after her bath and feeding. She is slipping into sleep. I yearn to join her, lying at her side. It seems right, yet it is not. I bathe and feed her as needed. I do the chores of a drone for his Queen. It is most appropriate that I serve her body's needs. I call her

Queen in my whispers and with the others. This evening I take it further. I unclothe her. I bathe her. I hold her naked body. Can it be true that I *am* her drone? Is such a thing possible?

My brother, my twin brother, goes to the Queen on Trania. He and I share the same genes. Yet, that day he is the choice, and I am not. I long ago blot any bitterness from my mind, but every now and then it resurrects itself. I am of the blood of the Drones. If not for fate, I mate with a Queen.

We grow up in hives, we Franian males. Most do not know which eggs they hatch from. We are mammals, but workers harvest the Queen's eggs and plant them like seeds. After time passes, we grow to become warriors. I never know my heritage, but when my brother goes to the Queen, I discover it. I am the spawn of a Queen. I am, therefore, a potential Queen's drone.

Yet, many times drones are not chosen for such a function. They do not spray the harvest of eggs with sperm, nor do they become the lovers of Queens. Those drones that do not meet the gene review are not full drones. My brother, my twin brother, is a full drone. Our genes are the same.

So, it goes around and around inside me. I grow acceptance. The gene pool cannot allow two identicals. Yet, if Prana is truly a Queen, that explains the rise of my scepter. I am a drone, after all. I am ready to mate with a Queen.

The thought shakes me, yet, no worse than those thoughts as I shampoo her slim, slippery body. To touch her, electrifies the currents inside me. I buzz with desire. Finally, I understand my body's reactions to her all these times.

Yet, I cannot give into these needs. Even if Prana is my Queen, she is still the High One of the RishMore, and she is an alien with strange thinking patterns and viewpoints. She does not understand a

Queen's path. She may never accept my lovemaking. She has made that clear. Yet, it is my job to serve her. I must do all that is needed.

I turn to glance back, then take another look at her. I breathe deeply, enjoying the scent I now savor. Her hair curls about her face. Before, my fingers entwine in it. They nourish its shiny lights. I sigh.

Her face, even gentler in repose, is smooth and sweet as morning dew upon the tongue. I yearn to caress her face, to once more touch the skin I feel before. Never could I imagine it feeling like that, so smooth — like liquid warmth upon my body.

I grow a bone in my pants. It throbs from the ache of my thoughts, but I do not move. I stand and watch the rise and fall of her breath. Beneath the covers, her creamy breasts lie against the sheet. In the pool, I see those breasts lifting up. Delicate curves of soft wonder. I close my eyes and recall their look and their smell.

It is said that to suck at a Queen's breast fills a Drone's heart with ecstasy. I understand that, as I stand watching my Queen. One of my hands reaches out. I am too far away to touch her, of course, yet, my mind plays tricks on the receptors of my fingertips, for in my mind, I am touching her breasts, holding them, caressing them. My hand rounds with the vision.

Her eyes open, and she looks at me. They are almost blue in the faint light of the room.

I gasp from the unexpectedness. "My Queen?" I ask. "Do you need something more?"

"Shnize, please come here," she says softly in a voice full of sleep and innocence.

The ache inside me grows more intense, but despite the pain, I move closer. How can I not? To smell something one desires is almost as good as to taste it. I say that over and over to myself as I sit down on the edge of her bed.

"I dreamed something strange, Shnize, even, maybe awful," she murmurs so quietly I have to bend down to hear her voice.

"I dreamed I really was a Franian Queen, and you were… I don't know what to call it. I didn't dream with words, but you were there, and you — oh, I can't say it. Never mind, Shnize, I…"

"You dream you are my Queen, and I am your Drone. Yes, Prana. I believe it is so. What else?"

"But I don't understand the dream. There was a Queen speaking to me. She welcomed me, and then other Queens gathered around me. And there were males who stood beside them. One of them looked just like you, but he wasn't you. His eyes weren't the same. He didn't smile at me."

"That is my brother, Prana, my twin brother. I believe your dream is real. Go on."

She sighs and pulls the blankets up, although I do not think she is cold. I place my hand on hers. She likes touch. It often reassures her. Yet, this time her eyes fly to my face, and they stare with fear and horror.

"Some of the Queens were very angry, Shnize. They were fighting about me."

I nod, "Yes, that is likely." I reach out and touch her face. "It is all right, Prana. They cannot hurt you. I will not let them."

She sighs again, but her sadness doesn't retreat. "Shnize, the Queen, the one with your brother, she was on my side. She spoke to the others. I don't remember what she said, but then, of course, I don't speak your language. I wish I did. I wish I knew what they said."

She is so like her young sons, fragile and full of trust. I scoop her body closer and hold her as I would one of them. "Do not worry about the Queens, Prana. The other Franians and I will always protect you. You know that. Do you not?"

She nods her head and lays it against my chest. "It's all silliness, Shnize. I'm sorry to bother you. It's just that it seemed so real. It was probably only from dealing with the attack, I'm sure. Nightmares are like that, a reaction to pressure."

I agree with her and hold her tighter. Then, I rock her small body back and forth. "Sleep," I tell her. "You are safe now. Sleep."

With her head against my chest, she slips back into sleep, and I lie against the wall, holding her, breathing in her scent, thinking over her dream. Perhaps I sit there like that for hours. I do not know, but as the night sends tendrils of itself into my limbs, I doze off.

I wake in the morning to hear her exclaim, "Shnize, you held me all night?"

I do not mean to stay throughout the night. I only intend to send her into deep sleep so she can rest well. I try to move out from under her, but she stops me with her hand.

"Thank you. I wouldn't have slept if you hadn't been here, Shnize. It was only because you held me that I didn't have more of those nightmares."

She reaches up then and kisses me. It isn't a kiss like the one I envision during the night, but it is sweet. I give her a quick hug and then release her. "Okay if I use your facilities?" I ask.

She laughs. "You mean you actually do urinate?"

I wake up with a stiffness down below that is much more than my need for letting go of water. I slide out of the bed and face away from Prana so she does not see. She is oblivious. She reaches out and touches my hand. "You're such a good friend, Shnize," she says. Then she closes her eyes, and I think, she drifts off again.

My urgency is great, but I can't help watching her a moment. Her eyelashes draw me. They are so finely threaded, dainty fragments of her hair, yet short. They are darker than the strands on the top of her head, where gold dust shines among the brown. Fleetingly, I wonder what color her hair down below is. I try to remember, but, the day before, I do not look, and besides water makes hair darker.

I cannot wait any longer; the need overtakes me, and I stride away. Seconds later, I groan from the release. The morning emptying is always a pleasurable sensation. I sigh. Then, I give myself relief in another way. Being so close to Prana leaves me no choice. I cannot lose myself in her scent and allow my Drone nature to overwhelm me. I do it for her safety.

I close my eyes and remember how she looks in the pool, her skin so silky and smooth, her young breasts only inches away from my hands. I relive that moment and think about her lying in bed, her body curled up in sleep.

I can go to her. I can run my hand over the soft texture of her skin. I can claim her. I think about doing so, working out the details, what I do first and then next. Ah, the sweetness of it. Release comes quickly and lasts several moments as I pump and pump my flow of hot, liquidy sperm.

When I finally leave the bathroom behind, I see that Prana is still sleeping. I sit down in a chair and watch her, trying not to think about my previous thoughts. Yet, her neck is open to my eyes. Her body lies there just as I imagine in the toilet room. She is so young, so tender. I know how soft her skin is to my touch.

Feeling my attention on her, Prana stretches, opens her eyes, and smiles at me. The sweetness of her lips and the tiny, white teeth take my breath. Despite my earlier precautions, once more the strength of my desire slams against me. I want her.

"Ready for breakfast?" I ask.

She groans. "Are you sure it's morning? Couldn't it still be the middle of the night? How do you know for certain that it's not the brink of dawn?"

"Time," I order. The computer comes awake immediately and voices the precise hour.

Then she cannot argue. She throws her legs over the side of the bed and sits up. "Did I really say I'd meet Jertot and Terriff at quarter cycle?" she sighs.

Her hair is messy, her eyes sleep-heavy, her lips slightly puffy from the way she's been lying. She looks beautiful. I yearn to tell her so, but I do not. I nod and add, "It is good that you meet with them so early because at half cycle eleven, you have Formalities."

She tilts her head. I see that she forgets, or perhaps, no one tells her. What person tells her such things? I know I do not.

"What is Formalities?" she asks, hesitantly, as if playing with the sound of it in her mind.

"That is when you come before your people for the first time. They salute you as their High One. There is no question in anyone's mind about your designation, of course, my Queen. Except with Terriff and his group of converts. I do not expect them to cause you any inconvenience today. They will gain no following by doing so, not while the ship is still singing your praises over the victory yesterday."

She sighs, not looking happy about the event. "But what should I wear? What should I do? I don't remember any lessons about this. Dar never thought I'd outlive him."

"Before, he cannot give you lessons, Prana. *You* form the Formality. I will show you a tape of his when we return to rest and change, but you must write your own path. As to your attire, there is no requirement. The dress you have on is suitable. Going naked is appropriate. You are free to choose."

"If those are my choices, you can be sure I'll wear a dress," she laughs.

I chuckle for a second, too, but her words remind me of her shyness in the water, and then it recalls to mind the way she looks and the way she feels in my arms.

I turn away to program our day's first meal.

Later, we eat and change our clothing, (I, of course, change into clean clothes as well. I select mine through the clothes machines.) As we pass through the exit door, I see that Frick is once more on duty. His eyes run a quick check over Prana, assessing her mood, appearance, and health. Then, content that all is well, his look turns sour.

"You accept Shnize as your chief Drone?" he asks the High One.

Prana doesn't know what he means. I think the visions she dreams in the night are now foggy in her mind. She turns her beautiful brownish-green eyes on me and asks, "Is that what Franians call you? Wait. I remember. The Drones are the ones who take care of the Queens. You *are* my Drone, aren't you?"

Prana, of course, does not understand the full implications, but it does not hurt Frick to think that I satisfy the High One in all ways. I nod.

Frick's eyes grow larger, then darken. "Is that wise? Are you capable? You have not been trained."

Prana is looking back and forth between us. I am sure she sees that the other guards are spellbound by the conversation. She blushes. "He doesn't mean…?"

"I take care of you, High One. I am, therefore, your Queen's Drone."

Neither she nor Frick asks any other questions after that. They both sense there is more, but, I think, they are each too embarrassed to probe. Actually, it amazes me that Frick engages his mistress in conversation. It is something he does not usually do.

We escort Prana to the War Room without problem. As usual, the corridors are vacant. The crew now avoid the main floors of the ship. I hear rumors that they believe the spirits of the dead haunt it. I am glad Prana does not hear that. She is too emotional as it is. If she believes Dar lurks in the hallways of the ship, I do not know what she will do.

As I start to enter the War Room, Frick grabs my arm, wanting me to speak with him. I do not like letting Prana out of my sight, but I mind-link with her and explain that I will join her shortly.

Her large eyes fly to my face. For a second, she panics. Then I see her steel herself. She reminds me of Dar as she does that. I am proud of her. I watch as she walks over to the table, her shoulders back, and her beautiful breasts jutting out into the air, saucy as her tongue. I wet my lips and turn back to Frick, listening to the door close behind me.

Chapter Seven

Prana

I realized at that point that I had become too dependent on Shnize, bereft just because he needed to speak with Frick. Yet, facing Terriff, a man who was not only my enemy, but who either wanted me dead or to drag me into slavery, not to mention killing my sons, was a bit overwhelming.

"Greetings to you, High One," both Captain Jertot and Terriff said. Stiffly, I accepted their welcome and sat. Terriff pounced into conversation at once, not even waiting for me to get comfortable in my seat. I supposed he was merely excited by the opportunity of speaking to me without Shnize. I attempted to listen.

"So," Terriff said after a winded speech about the possibility that Shnize was relaying our conversations to his fellow Franians and after a brief rebuttal by me which included an order to proceed with business, Terriff asked, "Have you come to a conclusion which plan to pursue?"

I folded my hands as Dar used to do and stalled for breathing time. "Yes," I said, after a quick check of Jertot's expression.

Terriff handed me a drink. Jertot had one, too, so I thought nothing of it. I took a sip and savored the cinnamon and chocolate warmth of it. "We will attack first on their right flank, hitting their government center," I said.

"And the spaceport?" Jertot asked, making it obvious what his opinion was.

I grinned. "That will be the second hit. I see no reason to be lax with these people. They deserve retaliation. It is obvious they are not about to give in peacefully."

"Bravo," Terriff said. "Well done. Let us celebrate. Let us drink to victory, High One."

"Put that cup down," Shnize ordered me from the door.

Without a word, I set it down on the shiny black surface of the table. Then I looked up, questioningly.

"I must always taste anything you wish to eat or drink, High One. Do you forget that rule?"

I paled. Of course. I knew that. Dar's son had tried to poison me once. Shnize had almost died because he'd received the dose. I knew he was my tester. How could I have forgotten?

"This is outrageous," Terriff cried out, standing up and glowering at Shnize, his nostrils flaring like a bull's.

"I did drink — but only a small sip," I said. "I forgot."

Shnize was at my side in a second. He picked up the cup and sniffed it. Then he glared at Terriff. "If you dare, or if you *ever* think to harm the High One, I tear strips off your skin, piece by piece until you resemble a raw, bleeding corpse."

"Shnize," I complained.

He looked down at me, shrugged, and took a gulp of the drink. Then he swallowed and studied the effect. A minute passed as he stood there. I sank down in my chair in misery, terrified that Shnize would suddenly succumb to the same spasms of pain he'd experienced before at the hand of a poisoner. I closed my eyes, unable to watch.

Shnize didn't collapse. Instead, he sat down beside me.

"It seems okay, High One," he said gently, "but I recommend that you drink or eat absolutely nothing given to you by the hand of *this* man. His loyalties are very much in doubt."

"This has gone too far," Terriff objected vehemently, thrusting himself up and out of his chair. "I refuse to hear any more of his insults. Get this Franian out of here."

I nodded at Shnize, not even bothering to look at Terriff. "I'm sorry I forgot," I told Shnize.

"Of course," Captain Jertot said. "I should have thought of that myself. You should not drink from any hand but your guard's. I am sure that is all Shnize meant, Terriff," he said, attempting to calm the situation.

I kept my head down, not agreeing or correcting. I suppose I should have soothed the War Lord, but I didn't want to. He was irritating me with his continued demands to have Shnize removed from the room.

"Silence," I ordered. "Sit down."

It amazed me that the man closed his mouth and took his seat. I wished I could have more of the delicious brew he'd given me, but Shnize was right. I'd been a fool.

Poor Jertot. His face was wrinkled into lines of worry. I opened my mouth to explain to him, but Shnize began filling him in on the situation.

"There is a suspicion that Terriff leads the rebels against the High One. The only reason the High One does not yet order his death is because she values his services as War Lord."

Terriff's face reddened. He slammed his fist down on the table, making me jump. "Guards," he yelled.

Frick came running in, followed by about ten others. "War Lord?" Frick asked after shooting a quick glance at me to check that I wasn't in need of assistance.

"Seize this Franian," Terriff demanded. "He has made vile accusations against me. He has…"

I stood and gestured. It was a High One's sign, one single clap and then two vertical hands, heading away from me and overlapping. I had never used the signal before, but I'd seen Dar use it several times. It stopped everybody cold.

"Hear me," I boomed out (Well, at least yelled louder than usual. My voice wasn't quite adequate for booming.) "War Lord Terriff has no authority in this matter. My Prime Minister Shnize is henceforth above seizure by ANYONE. He is my hands, my eyes, and my second brain. He is my Drone."

"Yes, High One," Frick said, bowing low.

"Your Drone?" Terriff turned to me. "What does that mean? You are a Franian Queen, now?"

I did not answer him. It was none of his business. My head remained erect and my hands at my sides.

"Head guard, answer me. Is the High One recognized as a Franian Queen?" Terriff demanded, glaring at Frick.

The man glanced at me, possibly wanting permission to give an answer. I saw no reason to deny his words. I nodded to him.

"Yes, War Minister. We recognize her as such."

Jertot moved forward. "What? How could that be? The High One is Terran, not Franian. I picked her up, myself. I saw her when she arrived. She is Terran."

Jertot glanced at me again, and then swallowed hard, realizing that he'd errored. "Forgive me, High One. I mean, you were Terran. Your heritage — that is all I meant. Before you became RishMorian, as you are now, I mean."

I smiled at him, not offended. Meanwhile, Frick waited for the captain to wind down. Then he, glancing first at me and then at Shnize said, "We recognize the High One as our Franian Queen. She is RishMorian. We do not argue that, but she is also our Queen, as well.

And yes, War Lord, it is true that every Franian onboard takes a vow to protect and honor the Queen and her sons. She and her spawn own our lives. She rules us."

Terriff's face grew red. His hands writhed, twisting like the Earth worm I had once seen in one of the Center's potted plants.

"That is ridiculous. Are you mad? Get out of here," Terriff growled. "Get all Franians out of my sight."

Terriff sagged down into his chair and sat with his arms crossed, scowling at us. He reminded me of a young child who has just been thwarted, his lip bulging with pout.

"Thank you, Frick," I said, giving the guard a smile of approval.

"By your leave?" he asked.

I waved him out, and the guards returned to their post at the door. Shnize, still sitting beside me, was sending mental messages so rapidly, I shook my head, unable to process them.

He sighed. "The High One is ready to continue battle planning."

Terriff glared at Shnize with the blackest scowl I'd ever seen. Then he shot a glance at Jertot, whose mouth was still at half-mast in surprise.

"But I don't understand," the good captain said as he moved back toward the table. "How can it be? What does it mean that you are Queen, High One?"

"It means that we protect her and her sons," Shnize said, eyeing Terriff with dark, dangerous eyes. "Each and every Franian strips the skin, layer by layer, off anyone who dares attack the High One or her sons. The blood coagulates, and the agony continues as the traitor screams and curses. For you see, we Franians do not have the sensibilities of RishMorians. We do not believe in easy deaths."

"Enough, Shnize. Terriff gets the picture. Can we go on now?" I said, shivering from the violence of my Drone's image.

Indeed, I wasn't the only one affected by the mental picture. Terriff had paled considerably. As we progressed, he became increasingly silent, broodingly so, and all discourse from him came only when someone asked a direct question. All the flirtatious, ego-boosting compliments he'd been lavishing on me dried up. I was pretty sure the matter of his ascendancy had been given a deathblow.

We were discussing our plans for the invasion after the bombing when Terriff took it up again. "You will, of course, take a Crezand husband," the War Lord said, interrupting something Jertot was saying. "Then you will spawn. May I ask, High One, how you plan to carry the seed of *each* planet we conquer?

"Have you thought of that? A man can easily produce heirs and still govern, but a woman's body labors beneath each new creation. She is spent early if she bears frequently. What will you do after your third or fourth alien seeding?"

I had not thought beyond Crezand. For one thing, I couldn't get used to the idea of mating with fly eyes. My fingernails dug into Shnize's arm, the one that I'd latched onto when we'd sat down and he'd started talking about peeling away people's skins.

"The High One is a Franian Queen, why is that difficult for her?" Shnize asked.

I gazed at him. I had no idea where he was going with that thought. I sat on the edge of my seat, eager to hear, but afraid I wasn't going to like it. Please may it not be something terrible, something even worse than mating with a Crezand, I said to myself.

"The doctor simply removes the eggs of the High One and semen from each alien mixes with an egg to achieve fertility, of course. A Queen need not bear her own young."

"But there have always been live births before," Jertot exclaimed.

"That is the procedure for most Franian Queens," Shnize told him, grinning at me. "An alternative is that foster mothers carry the fertilized eggs for the High One. Either way, she is far more important to RishMore as a leader than as a brooding body."

"I have a better solution," Terriff announced. "She will marry me. I will take her as first wife. I will honor her in all ways, even name her as my heir, as Dar did. Then I will plant my seed in her, and she will carry only my child, while I take the burden of being the conqueror."

It was a solution I might have gone for if Shnize had not explained to me what would happen to my sons. Terriff hadn't exactly proposed, but I took it as if he had. I shook my head at him. "No. Not so. Dar's sons would not live beyond the day you took me as your wife. Besides, I am the High One. I cannot discard that burden. My people would not allow it."

Jertot was shaking his head. "He would not kill your sons, High One. We would see to that. And it does make sense. Terriff is many years your senior, three hundred or so, is it not?"

"I have three hundred thirty-seven years of experience," Terriff informed us proudly, searching my eyes for interest in his proposition.

I shook my head again. Then I took a deep breath. I rose up and walked to the beverage machine. Shnize followed me, but made sure he arrived there before I did. He reached in front of me and said, "This is my job, High One. I shall always tend your needs. You have only to ask."

I realized I'd hurt his feelings. I was sorry, but I desperately wanted a drink. I still missed Terran coffee at times, but the RishMorian hot beverages were a close second. I wanted another sip of the brew that Shnize had taken away from me. Shnize picked up my thought and spoke the order.

Terriff, too, had followed. He stood so near, I could almost feel his heartbeat. "There is another matter, High One," he said as Shnize picked up the mug. "It may be indelicate to bring this up so soon after the death of your husband, but I must voice my concern."

I glanced at Terriff, surprised that he'd so quickly regained the bravado of his previous manner. His posture was again the cocky, self-assured nobleman as he towered above me, leaning slightly into my face, much closer than Shnize liked.

Shnize waved him backward, but Terriff was looking at me and either didn't see the gesture or ignored it. The same handsome smile was back on the War Lord's face, and he reeked of confidence.

"You are without a lover, High One," he told me, speaking in a whisper loud enough to carry clear across the chamber. "You need a man, *Prana*. I have had many years to learn the things that please a woman."

"Stop it," I hissed.

"Dar was an excellent lover, was he not? I would be his equal," Terriff whispered, moving even closer.

I shuddered and backed away. Terriff's offer was absolutely revolting. How dare he bring my husband's name into such a proposal.

"I don't need a lover," I told him, shaking my head.

"High One," Shnize interrupted.

"What!" I yelled at him. Shnize was holding out a mug of the badly needed brew. I stretched out my hand to take it.

"Sorry, Shnize," I said, taking it from him. "I didn't mean to…"

"Do not apologize to your guard, High One. He is merely a Franian," Terriff said in a voice even louder than his whisper.

I glared at him. "He is my Prime Minister. Have you forgotten that? Besides, I like Franians. I like them better than *some* RishMorians."

Shnize shook his head at me. I took a sip of the drink and sighed. Being nice to Terriff was getting harder and harder.

"Besides," I said, raising my eyes to stare into Terriff's. "I shall soon have a new husband."

"The Crezand? You will endure his touch? You are a brave woman. I like that. I will wait for you to tire of servants and big-eyed monsters."

"It would be better if the High One retired now," Shnize broke in suddenly with a gruff and overly loud voice. "At half cycle eleven she has the Formality. She needs to rest."

I took another long drink of my perfect temperature brew, walked over to the table and reluctantly set it down. "You're right, Shnize. Thank you for reminding me."

I glanced at the captain and at the War Lord. The latter's ears were flopping about like a wounded bird. His eyes were like darkened storm warnings. My rejection hadn't injured his pride. It had only made him angry.

"I bid you good day, War Lord Terriff and Captain Jertot," I announced. Then I turned to leave.

"Wait. Please," Captain Jertot said, standing up. "High One, I never meant that I doubted your right to rule. I was only…"

He gathered his breath and started again. "Dar had faith in you, High One," Jertot continued. "I saw that from the first. He repeatedly spoke of your sharp mind and your amazing abilities. He loved you, High One, and I know you felt the same. I saw it in your eyes.

"For that reason, above all others, I have not spoken to you about… But since Terriff has, I see I must. Perhaps, you see, Terriff may have put the idea into your mind that you need… It is the same idea that I have had, High One. I mean no disrespect when I say — Forgive me, Shnize — but I offer my services in the way that a Franian servant cannot.

"I can give you two hundred thirteen years of experience, my dear. But should you prefer a man who would not wish for power, a man who would never want more between us than you allowed. I am willing."

"I would be that man," Terriff interjected.

The captain lowered his head into a respectful bow and completely ignored the War Lords interjection. "That is all I wished to say, High One, except that it would be an honor, if you ever — if you found me acceptable," Jertot continued disregarding Terriff's high flying arm, which indicated that he'd taken umbrage to the offer.

My eyes teared. I knew that this speech had been difficult for Captain Jertot. He was not an outgoing man. He had derived his position only because Dar had placed him there, but I had already decided that he was a friend, a good man, and a reliable one.

I nodded. "Thank you. I would be much more inclined to place a bite of food in your mouth than in War Lord Terriff's. (In RishMore, exchanging food is a prelude to a marriage agreement.) You are honest, Captain. I appreciate that. But I grieve for Dar. I cannot take another in my husband's place. I am still too full of him."

Perhaps it was not wise to compare the two men so openly, but I wanted it clear to Terriff that since I couldn't trust him, I would never accept him. His body and mind called to me, but to have him in my bed would be like making love to a viper.

Shnize took my elbow, something of late, he'd often done. I suppose to him it seemed undemonstrative, or perhaps it was a thing a Drone would do. I moved forward as I felt his body take the first step. Mentally, I was still shaking my head at the conversation of the past few minutes. Was it always like this for a leader? Would everyone proposition me in the future?

We had only just stepped through the doors and were heading through the corridor when Shnize spoke, startling me from my thoughts. *Do you desire Captain Jertot?* Shnize asked, sending the question into my mind with a mental force that jarred unpleasantly.

I gasped and stopped.

"Sorry," Shnize said, urging me on, but the thought still dangled there unanswered. I was musing about Shnize's reason for asking it when he repeated the question. I glanced at him, seeing only his usual mask of indifference. *No, I don't desire him. I don't desire anyone. But I do like Jertot, though. He's a good RishMorian.*

Shnize let out one of those noncommittal grunts he made now and then, a grunt that didn't seem to have any meaning, at least not one he ever cared to explain to me.

People were abundant in the halls that day, and they were all heading in the same direction. When they saw me, they stopped, bowed low, and waited for me to walk past. I accepted their obeisance with a quick bob of my head, copying from my memories of Dar. No one spoke, nor did I, but the RishMorians stared at me, studying me as if I were new to the ship.

"The crew assemble now. All is ready," Shnize said. "We restore the Great Hall enough for today's Formalities."

I remembered the room vividly. It was where I'd long ago married Dar. I had not been back since, although it was where Dar had gone to wed Mada, the Crezand who'd murdered him. With that thought, I stumbled. Shnize's grip on my elbow tightened. *Do not be uneasy,* he said. *The Formalities go smoothly.*

I shook my head. *It's not that, Shnize. I was just remembering about my wedding and Mada's and how soon I must endure that spectacle, too. Will a husband be as weepy as a wife? Or will he accept me? I won't need to drug him, will I?*

Shnize smiled down at me. *How could a man — any man — be other than eager to bed you?*

Of course, I blushed. My face grew hotter than molten lava.

***

For a while, when we returned to my room, I just relaxed, but I was becoming more and more agitated about not knowing what to do during the Formalities.

Shnize brought up the vids of Dar's first time being accepted as ruler. He looked magnificent, poised and easy with his fame. He was so much younger than when I'd known him. I sighed and watched the vid over and over until Shnize finally growled at me. "Are you learning what you need to learn or just marveling at the beauty of your husband?"

I laughed, but Shnize was right. Dar had been so incredibly handsome — once I'd gotten used to his alienness. Seeing the younger version of him was even more breathtaking.

"Okay," I giggled. "Once more. This time, I'll try to concentrate on the Formality," I told Shnize, but it was hard. Dar was Dar. He commanded attention, even in a vid.

"Was he kinder then, more gentle?" I asked Shnize.

He laughed. "Kind? Not at all. The youthful Dar is even colder than when you know him. He weighs everything objectively. He does not fall in love with a girl from an insignificant planet. If he bothers with Earth at all then, he would marry you and give you only a brief training period. Then he would send you back, expecting you either to survive or not. If the After comes for you, he would not care. He simply would take another. At that time, he lacks all feelings."

I believed what Shnize told me, but it was difficult to conceive.

"In his early life, he is an impatient man," Shnize added. "Quick to judge, fast to condemn. Your flashes of rage he would deal with harshly. When he is young, he would cripple a wife who runs from him. I am glad he is older when he meets you."

I sighed and looked away from the image on the wall.

"You cannot love the man you see in that vid, Prana. I think you would only know strong hate."

I nodded that I understood. I should have hated Dar, anyway. I don't know why I didn't. He'd conquered my world. He'd twisted my heart."

Yet, oddly, I still missed him. His absence was like a hole that kept expanding day by day into a bigger one. Soon, I guessed it would eat me up. Then I would no longer exist.

Shnize either read my thought or saw it in my face. He came over and sat down beside me on the bed. He handed me a cold drink and a fruit that he'd just ordered for me. "Eat something," he said, softly. "You need your strength."

"I'm not hungry," I responded, pushing it away.

"Then you can feed me," Shnize said, making his first joke.

I stared at him in shock. Then I realized exactly what he'd asked, what it meant, and the way he'd said it.

His mind was not bombarding me with anything, but his eyes were sending messages. It was a joke, wasn't it? Or did he mean it? My heart, for some reason, began to chug-a-lug much faster. My mouth went dry. My blood — let's just say, that I didn't feel the way I'd felt when Terriff had suggested the same thing. And when Jertot had offered, I'd felt nothing. But now…

But Shnize was a Franian. He couldn't. I sighed. Then I shook my head. "Weren't Terriff and Jertot enough for one day? Now you're starting in?"

"I only want to relax you, Prana. I will not hurt you. I promise. Permit me to take care of you. Please."

I didn't say anything. My voice was too unsteady, but I turned off the vid, no longer wishing to stare at the younger vision of my husband. My heart hurt from staring at him, anyway. And what Jertot, Terriff, and Shnize were all saying was true. I did need someone. I needed Dar, Dar with his arms around me, his body next to mine, his…

"I can't. I just can't. It wouldn't be the same," I told Shnize. "You're not…"

Shnize stiffened. "I am not Dar, but I am your Drone. Feed me or not, I worship your body. I …"

"No," I said, cutting him off with a finger on his lips. "I can't. You're right; I need someone. But I have to love him."

"Can you love the Crezand you take as your husband?" Shnize demanded, his anger flashing into the darkness of his eyes — like the blinking city lights I'd long ago seen from my hotel window.

"Stop it," I told him, looking down at my twisting hands. "I told you I can't."

Shnize's hands reached out and gripped me. "I obey you, Prana, but you are wrong. You follow my advice with every other thing, yet you discard what I tell you about your need. Why?"

I couldn't answer that; I just started to cry. Shnize was unbalancing me. I didn't know what I wanted anymore. I did want him, but I couldn't do what he wanted. Not that way. I needed a man who could…

"Look at me, Prana. Now — Prana."

I was trembling. His scent was in my nostrils. It was making my head swim, but I obeyed. I opened my eyes and looked at him.

"Good girl. Now, let me just kiss you. Try that and see, Prana. I stop if you command it, but permit me to try, my Queen, please."

I could feel his mind waves. I knew what he was doing. His thoughts were strong, urgent, demanding, yet it wasn't the pressure of his mind which made me nod my head. It was because I was curious what it would be like to meet his lips. The truth is I wanted to kiss him.

I said nothing, but he must have read agreement in my eyes or from my mind. He bent forward, and his lips touched mine. I didn't pull away or forbid it. Then he wrapped his arms around me. He had done that many times. I felt secure in his hold. I sighed, almost contentedly and waited to see what else he would do.

My husband Dar, at that point, would have attacked. His lips would have forced mine apart. His tongue would have thrust itself inside me, always the conqueror. Shnize didn't do that. He moved gently and slowly. His lips didn't join mine after that first peck. They traveled down to my neck.

I don't know why I whimpered. I didn't even believe I had. I thought the sound belonged to someone else, but there was no one in the room. The warmth of Shnize's breath on my neck, the touch of his lips made me cling. It made me breathless. Again I moaned, and then I panted.

"Shnize," I said, which spurred his kisses, the touch of his lips, and his fingers caressing my face. Everywhere his lips touched, my skin desired.

Slowly his intensity rose higher, raining a line of sweetness up my neck, my chin, and teasingly just beneath my lips. My heart panged like it had been swept into restless currents, my blood became the white waters of a river. It rushed here and there, urging speed, racing to a far-off waterfall of pleasure.

Then, finally, Shnize's lips touched mine again. At the same time, his hands moved behind my neck, and he pushed me more closely into an arrangement as old as RishMore. I opened my lips and took him in. I don't recall doing so, but it happened so easily, so quickly.

The taste of him was not like Dar. His body didn't smell the same way either, but it felt right, and it suddenly smelled as agreeable as a delicious dish. I opened to his touch. Our kiss lingered on and on as we got to know each other's taste and feel.

One of Shnize's hands moved downward. Wherever he touched my skin grew hot, and I shivered from the strange, desirous ache. I wanted Shnize's hand to touch me everywhere. I wanted so much more. And then the hand slid onto first base, comfortably, rightfully onto my breast. I wiggled with contentment as his hand and fingers cupped and squeezed and caressed.

He pulled away from me and looked down into my eyes. "Do you command me to stop now, my Queen?" he asked.

His fingers had switched to massaging around and around, stroking my other nipple. I was limp, wet, and ready. I slid my hands around his neck and pulled him back to me. I thrust my tongue deep into his mouth, urging him on without words.

One of his hands continued his play at my nipple, but the other kept moving down. I should have stopped him, but I wanted that hand. I needed it to touch me. Shnize continued, and without stopping, slid into home base. Cupping my most tender area, he shifted me slightly. I stilled my tongue, freezing for the moment, but he didn't. His hand probed down inside me, into my inner depth.

He was trained, yet untrained, if that made any sense. Shnize had told me that himself, but his inexperience, in a way, frustrated me roguishly. He flitted about, uncertain. I wiggled, but he didn't know where I needed him to be.

"Shnize," I groaned.

His mind touched mine and sucked in my desire and built his understanding. Then he knew what I needed, and he pleased me. My body rocked with passion, my mind melded into his, my heart stopped, and I gasped with ricochets of delight.

He waited a moment, but he wasn't through. Franians were neuters I'd always been told. I'd thought that Shnize's fingers would please me into fruition as they'd done, and then there'd be no more, but he tore off my clothes and his, and I saw that he was as ready as I for more.

"But you're Franian," I exclaimed, in disbelief.

He laughed, a laugh so deep and raspy that waves of desire flooded me. "Yes," he said. "I am Franian, but I am a Drone, my lovely Queen. I can ride you until you touch the sky and go far, far beyond."

Thus, he slid into me, and the pitch of our desire took us upward into a dance among the stars and their revolving planets.

When we came back down, and our breathing evened out, Shnize continued just as sweet. His kisses eased my racing heart. Then his lips on my nipple promised many more pleasures to come.

"You are flower nectar, skylights, and the taste of ecstasy. Thank you, my Queen."

I hugged his hard, blue muscled body, and kissed his chest. I sighed in contentment. "You were so right, Shnize. I did need you. I didn't know it. I had no idea it would be like that. I didn't understand."

He laughed. I wondered if it was a habit he'd picked up from me. "I know it is like that, my Queen. I imagine it many, many times." His lips once more joined mine, and his hand slipped back into place at my breast. Life was good.

"It is time to get ready," he said with a gentle squeeze at my nipple. "Even though I prefer to stay and ride my Queen again, and I think you are ready for another round," he said, stroking both of my breasts, "but, you must prepare for the ceremony."

I wrapped my legs about him and refused to move out of the bed, but he simply lifted me up and carried me to the pool. We swam there together for a moment, playing and kissing and touching body to body. It was painful to get out, but Shnize was insistent. Slowly, languidly, we rose like a god and goddess stepping out of a pristine lake, each naked as the trees and the animals of nature.

We dressed even more slowly, our hands playfully helping each other with garments that seemed ridiculous to don when our bodies were so much happier without them. But at last we were ready, Shnize in my color, and I in a long, full-trained dress in the richest emerald velvet, a fabric so soft it made the vest Terriff had worn the day before seem harsh to the touch.

"I feel like I'm floating," I told Shnize, and once more he laughed.

"You look full of contentment, which is appropriate. Your people will see you at your best. Look in the mirror, my Queen. See your eyes. They twinkle with highlights. They are sparkling orbs. Your cheeks flush half scarlet. Your lips are full and wide as two lappa petals, and they still pulse from the heat of your desire. You are beautiful, my Queen, and everyone on board soon sees that. They can no longer wonder that Dar chooses you. They witness and understand why."

"But I thought that was what I didn't want them to think, that Dar chose me as High One just because he enjoyed bedding me," I giggled.

For the first time, Shnize threw his arms around me and crushed me against him. "It is good that the people observe you like this, my

Queen. They know you choose no one. They begin to believe that the High One, himself, still visits your chamber and fills you with delight. They assume that Dar still rules from your eyes."

"What?" I twisted around, out of his grasp so I could look at him. "I don't understand. Why would they think that? Dar is in the After. How could he come to my room?"

Chapter Eight

Shnize

I remember then that I shelter her from this knowledge. I do not mean to divulge it. Yet, it slips out, and so I must explain to her. I hope that she does not find it too alarming.

I watch her eyes as I feed her this knowledge. I do that throughout our lovemaking. Her eyes always speak to me — even when I cannot read something from her mind. Her smiles tell me certain things, and the movement of her body displays her passions and her moods, but it is her eyes that help me most to understand her.

"It is not true, of course, my Queen," I say, "but some members of the crew believe those in the After walk among the corridors. They think they detect strange lights and hear noises that cannot be. It is only repairs in the making, but some people are very superstitious. They suppose that Dar walks among us still."

"I hope not," I say. "He would be very angry about what we just did."

"Are you sure?" I ask, holding her away from me, studying her eyes, reading what her thoughts say.

"He was always jealous," she tells me, cautiously, trying to avoid looking at me. "He would never permit me even to be friendly with you. Remember?"

I smile to relieve the tension. I see she still worries about what he thinks. Does she feel guilty about our actions? I cannot allow that to happen. I know I need to talk her out of it. Yet, we are late already.

"In the past you worry Dar with your egalitarian ideas," I explain to her. "He is always aware of position. But, Prana, he knows that I love you. He reads that from my mind several days after you arrive."

Her eyes enlarge. The pupils grow round with her fear. "He knew that? And he still let you guard me? Why? That's not like him. I don't understand."

I smile down at her again. Then, I run my finger across her cheek, watching as her body registers my touch. Dar once tells me that Prana is the most receptive of all his wives.

"He trusts me, Prana," I reassure her. "He reads my mind, and he sees that I do nothing to cause you harm. He knows I do not risk frightening you by telling you my heart. You are safe with me. Dar holds certainty of that."

She nods. Everyone knows that Dar probed those around him, picking up even stray thoughts.

"But I can't believe he still allowed you to be my guard," she repeats, her expression full of wonder. Then her eyes widen even larger, and she looks up at me with tears in her eyes. "That day you shampooed my hair. You wanted me then, didn't you? But you never…"

I stop her with a caress. Then, I bring her hand to my lips and kiss each finger. "Then, you are not ready, Prana. You are full with fear. How can I burden you with my feelings? You need reassurance. You are almost not ready this time. I worry that I move too fast for you."

She blushes and looks down. "I never knew. How could I have been so stupid?"

We are very, very late. It is past time to leave. I give Prana's fingertips one last kiss, and then I retake my hold on her elbow. It is the only public touch I allow. I resent that. I want to hold her tight, to scoop her up against my chest, to feel her heart beating with its odd, funny, little rhythm.

I wave my hand, and the door slides open. Frick is ready for us. He clicks his heels into attention, bows briefly, and does his usual span of the High One.

"Meruft," he says. "Rethrun porthre cretrow. Merrowth," which loosely translates to: Expletive; you satisfy the Queen well. She glows."

I give him an equally quick bow of thanks for his praise and run my eyes over the others. They, too, bow low to me. They approve and validate my new status.

"What did Frick say?" Prana asks.

I steer her forward. We need to walk fast to get there before Terriff commits some foul and treacherous deed. "Frick says you look beautiful, my Queen."

Beside me, Prana is stretching out her much shorter legs. Her lips part, and she is keeping up with me but panting slightly. I slow. It is not seemly to have the High One arrive out of breath.

As we walk, I notice that the sparkle of Prana's eyes and the ruby in her cheeks are even more obvious from the excitement and from her sprint. It pleases me.

A force of Franians joins us as we near the Great Hall. They bow to their Queen. Their chief gives me an eye and then formally nods to me as well, giving acceptance to my new position. Our love making is that obvious to my Franian brothers.

We hear the roar of the crowds as we approach. The ship holds close to a thousand RishMorians. Not all of them will come, but a good majority of them will. I whisper this to Prana. She nods. She remembers from her wedding. She has a good idea what to expect.

The guards throw open the doors. No one announces her, but as Prana walks in, her head high, her poise every bit as regal as Dar's, the crowd silences.

Almost immediately, it is obvious to me that Prana grasps more from the vid than just the handsomeness of her husband. She flows through Formalities with the talent of a natural leader. The people's faces absorb her as flowers take in the sun.

I remain by Prana's side throughout the ceremony. Occasionally, she glances at me, gives me a quick smile, and then turns back to her people. I am proud of her. I hope having power will not change her. I worry about that, remembering how cocky Dar grows to be after his ceremony. But Prana is special. I secretly slap my hand for good luck.

Captain Jertot is first at the speaker's podium. He introduces Prana, or rather confirms that she is the heir. Several nobles then step forward, pledging their loyalty. Terriff is not among them.

When there is a lull, Prana introduces me. That stirs up the crowd. No one introduces a Franian, nor gives one a title. But it is Prana's right. That is not open for query. Except it is. Terriff strides forward and takes the other speaker's position.

"What is this?" he demands, frowning out over the crowd. "A High One who needs to lean on a Franian?"

"I honor Shnize's service — as did my husband, Dar, before the After," Prana says without a pause.

Many of the people nod. Everyone knows that for centuries I am constantly at Dar's side. But there are some who shake their heads. Several people even jeer. It is not clear which speaker they mock, the High One or Terriff. I do not think they dare boo the High One.

Unfortunately, the War Lord is not finished. "I say that a Franian deserves no title," Terriff continues.

"You question the position of the High One, Upstart?" Prana asks with just the right touch of affront and insult. She lays it on rather heavily with the slur, but I almost smile from its result. Terriff drops his act of composure and beats the podium with a fist. That belittles him in the eyes of the crowd. The front row hisses. This time it is clear who loses favor with the people.

Apparently unaware, Terriff continues. "How dare you call me an upstart. I inherited my position, just as you did."

"Thank you for your recognition, Terriff," Prana says, with one of her beautiful smiles. The crowd roars. So saying, Prana traps Terriff into publicly recognizing her as the High One.

Terriff stutters, "I…I…I…" The crowd cheers, overruling whatever he might say next. Their boisterous "Umpha! Umpha! Umpha!" goes on for several minutes.

Terriff glares at the people and then at Prana, but he does not give up yet, although wisdom demands him to cease his challenge. The crowd is in love with Prana. They adorn her with fresh cheers.

"This woman was merely the wife of Dar. She is too young," Terriff yells, giving way to temper once more. He repeats his words, over and over, screaming them out into the crowd until the RishMorians grow silent, watching him with distrustful eyes.

"Too young?" Prana says, without any noticeable loss of composure. "Our former High One, Dar, my husband, did not believe so. It was he who named me Exalted Leader of the RishMore."

The crowd loves it. Again they flare into cheers and shouts. "Umpha! Umpha! Umpha! Umpha! Umpha! Umpha!"

Prana surveys the crowd, her eyes traveling from face to face. She cocks one of her crooked little smiles, and the cheers grow even louder. If at that point she demands the War Lord's head, they give it to her. She can name anything. They treasure her.

Foolishly, Terriff continues. "Take a good look at your Exalted Leader. Do you not question why her face glows with contentment, yet her husband is dead? She has been fulfilled by a Franian, the same slave she wants you to bow to."

The crowd breaks into laughter. The RishMorians believe that Franians cannot service a woman. Terriff loses even more points.

Prana glances at me. Her poise, perfect up to that moment, bleeds into uncertainty. The attack startles her.

It's all right, Prana. Tell them that I will speak, I mind-send to her.

The Queen's eyes hold the first hint of doubt.

Remember your sons, I remind her. She nods, raises her hand for silence, and speaks with a voice that sounds just as confident as before.

"My Prime Minister will speak to that," she says, and then she looks at me with such trust and confidence, I know I cannot let her down.

"Citizens of RishMore." I say, and they quiet, curious to hear what a Franian chooses to say to them.

"For many centuries, I am the servant of Dar, the High One to the RishMore. He gives me to his wife so that I can guard her with my life. I do so, and continue to guard her with diligence. Anyone who ever attempts to harm her by word, by deed, or by other injury bears my wrath eternally.

"However, it is not for this reason that the High One, Prana, bestows this title on me, but because for the hundreds of years that I serve our former High One, I observe and record all that occurs. It is that information which makes me of special value to my High Mistress. She wishes my recall of events, my observations, and my recollections, all of which, of course, I offer freely. As all Franians, I live only to serve."

"And do your serve the High One in the *bedroom*?" demands Terriff, interrupting me.

The boos that issue forth then are such a thunderous wave of negativity that even Terriff, who so far is unaware of them, recoils.

I frown, wait for the room to be silent again, and then say, "I serve my High Mistress in whatever way I am *capable* of, War Lord. But you, yourself, do you not remember the words at your proposal to her this morning — and her rejection…"

The RishMorians gasp. Faces grow irritated, and some, even hint of the sternness of retribution. Terriff is no longer one they can respect. Prana is in mourning. Tarriff's premature proposal angers them. I wait for their cries of outrage to cease before continuing.

"Do you not clearly remember her words? I do. The High One coldly protests that she still grieves for her husband. I remember her sentence. Her words: 'It is only his arms I desire.' Is that something you still choose to forget?"

Terriff ignores me. He repeats his former allegation. "You can look at the woman and see that she has been …"

He breaks off when the shouts of the people out voice him. "Dar lives. Dar lives. Dar lives," they are saying.

Terriff blanches, turns to look at me and then glares at Prana. "All right. You win," he says. *For now*, he mind speaks the two words across the room so skillfully that I hear Prana gasp, and the hollow shout of his voice painfully echoes inside my head.

"I accept the High One," the War Lord says gruffly and stomps away. His departure leaves the room ready for the string of others who take their place at the podium so they can give their blessings to Prana's succession. These speeches continue for several hours. Prana begins to sag. She leans against the podium, allowing it to bolster her. More than once, I speak sharply to her using the mind link, but she is not a male. She does not have our endurance.

Finally, the last of the nobles and of the officers under her command pass, each vowing to follow her. Then Prana straightens up and speaks to the crowd. "Ah," she says, smiling with such sweetness and beauty that they, too, fall in love with her. "I thank all the nobles, but, like you, I am delighted that now the party can begin."

She couldn't have said anything more fitting. The people throw up their arms and chant: "Umpha! Umpha! Umpha! Umpha! Umpha! Umpha!" Then one lone crewmember brazenly adds, "Umpha Stegana," which loosely translates "Hurrah for the beautiful one."

Across the entire front row, they take up the new chant, "Umpha Stegana." Then it spreads outward and the entire Great Hall embraces it, "Umpha Stegana, umpha Stegana, umpha Stegana.

Prana leaves the speaker's stand and walks down into the crowd. I, of course, accompany her, as does her private guard team with Frick in command, but it is still a brave thing she does. The crowd draws back and makes a channel for her.

It is a crime to touch the High One. The crowd is well aware of that. But no one wants to harm her. They wish to share her breath, to inhale her lovely fragrance. Their heads bow, their eyes hit the ground, but as she passes through their midst, when their heads bob back up, they are smiling.

Prana walks toward the tables. She stays a while, pretending that she is tasting all the food, but she actually eats nothing. I see to that. Then at last, the time comes when she can wave a formal good-bye, thank them all for coming, and leave so they can enjoy their feast and dance.

As she departs again, the crowd erupts into "Umpha Stegana." Prana walks with perfect grace through the grand entrance and continues far into the next corridor. Then she stops and says, "I need to sit down. I am just too tired to make it back to my chamber."

I smile, sweep her up in my arms and carry her the rest of the way. Her arms wrap around my neck, and she sighs contentedly. "I was hinting," she says with a darling smile.

I want to stop and kiss away her fatigue, but I am still on guard. I do not trust Terriff to give up so easily. So I carry Prana stiffly, my eyes on patrol. I notice that the private guard team with Frick in the lead also remains on high alert. That pleases me.

"Well done," I tell the guards at the entrance to Prana's chamber."

Frick surprises me. He not only accepts the compliment but responds, "Thank you, Queen's Drone and Prime Minister," and then he bows low and respectfully.

Terriff

She made a fool of me. That small, alien woman, a female no bigger than a child. She made a fool of the RishMore's War Lord. It burns my insides as I sit here dousing disappointment with liquor.

My crystal goblet is almost empty again. I clink the cup against the table and demand more. The robot slides back to my seat. "Another?" it asks.

"Haven't you had enough, Terriff?" Sandoff asks, attempting to take my glass.

I glare at him. "You did not see the look she gave me when she rejected me this morning. She hates me."

"Yes," he nodded. "I can see that she has a good reason. You wish her sons dead. I would think that was sufficient for the loathing of most women."

"But I offered the woman marriage. I told her she could be my first wife," I explained, only slurring a couple of the words as I spoke.

Sandoff clapped me on the back. "Maybe she will still come around, my friend. She is only a woman, after all."

I gulped from the filled goblet. "Did you see her today?" I asked him drinking down at least half of its contents. "She is a true Queen. That is what the Franians call her, a Queen. Do you know what that means?"

"It means that we better go slowly," Sandoff growled. "She is even more deadly now, armed with a race that has infiltrated every ship, every planet, and waits attendance to every lord of the realm."

"But women have always found me charismatic. Why does she not? Why?" I asked, downing the rest of the new drink. I tried to call out for the robot to bring me another, but Sandoff was suddenly hoisting me up.

"What are you doing?" I demanded. "Is it time to go?"

"Yeah, time to go, War Lord. Tomorrow is already here. Do you not have an appointment with the High One at quarter cycle six?"

"Yeah," I said laughing. "That is good. Another chance. I can try again. Right? I will smile at her. I will send my mind into hers, and she will yield to me this time. But give me another drink, first. Robot, robot!"

I was still calling out for the idiot thing as Sandoff dragged me out the door. He propped me up against the wall since I was a bit unsteady. Then he took off in the other direction. I wondered where he was heading, but my head suddenly felt velvet-stuffed, and I couldn't worry about that. My guard helped me stand straighter.

"Where to, War Lord?" the Franian asked, but I ignored him as best I could.

"She will still be mine," I called after Sandoff. Then I slid slowly down the wall. The last I knew that idiot guard was throwing me over his shoulder and carting me off to my chamber.

"Your Queen is going to be mine," I told him.

The guard did not answer. He should have said something. He should have agreed with me. I started to speak to him about his insurgence, but I guess I dosed a minute. I don't remember anything after that.

I woke up the next morning in my bed. Some blasted guard had opened the door and reminded me of my appointment with the High One. I stared at him. I didn't remember if it was the same guard as the night before. All Franians looked alike to me. I made a rude gesture at him, ordered him out, and went back to sleep.

I was very lucky. The High One, herself, canceled the meeting. It was the first thing I learned when I finally woke up enough to order my messages read.

The relief was a slap in the face as I thought about what could have happened to me. But then I began to ponder the cancellation. Why had she disregarded the meeting? What was occupying her time? My heart sped up, and rage built inside me as I pictured her with that ugly, short-eared Franian.

Growling, I threw on my clothes, and rushed through the door, and sped down the corridors toward her chamber.

Prana

The Formalities had taxed my energy. I felt drained when I woke in the morning. I rolled over and found Shnize awake, staring at me with a gentle, kindly expression.

"Do I have to go to that appointment today?" I asked. "I wish I could play hooky. I'm tired after yesterday."

"I do not know the game 'hooky', my Queen, but I shall learn it if you wish," Shnize told me, flashing a smile and bending forward to kiss my lips.

I giggled. "'Hooky' means 'not going to school.' It's because I don't feel like doing business today. I need a day off."

Shnize sat up. "Then you cancel your meeting. After yesterday's success, you deserve to play this 'hooky,' my Queen."

I smiled. Then I jumped out of bed, ran into the bathroom, and performed my early morning necessities. I stretched leisurely.

"I want to visit my sons, today," I told Shnize the moment I came out. He nodded and served me breakfast. That was not all he served me with that morning. It was much later when we headed through the corridors to go see the boys.

That is why, when Terriff raged at the door of my chamber, I was not there. According to the accounts of the Franian guards who escorted him, the man banged and thundered for quite some time, demanding that I open the door to him. I'm glad I wasn't present. That way I can giggle over his impertinence.

I have decided not to comment over it. I was told Terriff had been drinking late into the night. Perhaps when he recalls with lucidity his demand of the ruler of all RishMore and what I could have done to him for it, his behavior will improve. I can only hope so. He is a good War Lord.

***

We arrived at Crezand's outer proximity a few days after the Formality. I stared down at the planet and wondered why this one of all the others had been successful with their tiny insurrection. We still didn't know how Mada had carried the bomb past all the screenings.

What ingredient had she used? What was the secret weapon? The scientists onboard the ship had still not come up with the answers, although they'd run the fragments through a spectrum analysis over and over. Whatever it was, its traces were destroyed in the heat of the explosion.

I have visited the site of the disaster several times, watching its restoration. It has been progressing well. The dining hall should be ready shortly for reuse. I've had a plaque put up. The names of all the dead have been carved on it. It was the only tribute that we could make to all we lost. The RishMore hold no services for those who go to the After. They simply mourn them in silence. Even my plaque is pushing Terran customs on the RishMore, but I demanded it. I wanted the mark of respect for Dar and the others, a reminder for each time we passed the scene.

But it was Crezand I was talking about. Its grays and browns loomed there below us, which meant that the day for war had arrived. It was the day that I said the words that sent disaster down upon so many.

Did I strike at the ones who had caused Dar and the others' deaths? That I'd probably never know. Perhaps I brought destruction down on innocents, people who had no idea what had occurred. They didn't deserve their pseudo deaths, but it was done, as it must be. To strike at the RishMore was cause for automatic retribution.

The order was easier to give than the penalty of knowing what I'd done. Terriff made sure I saw the close-ups that came after. They were displayed with the names and the pseudo death pictures of each individual. I endured the stream of images because it was my duty. In the Weapons Room, my face never quivered with tears, never flinched at the sight of the Vids of weeping Crezands or the faces frozen into long-term punishment. Crezand would see those faces, statue-like for centuries. They would live with their past and the warning for their future.

As I said, I was strong throughout the bombings, during the follow-ups, and I endured Terriff's taunts. I was strong until that night when I cried my grief in Shnize's arms. Then the faces flashed before me, the eyes so different from Terran or RishMore, but with tears as our common link. Mothers, fathers, and children. Real people. I sobbed and hated being who I was. Shnize got me through that night. He warmed my icy heart and restored my sanity with his gentle words.

Terriff

The High One never mentioned the day I assaulted her door. Perhaps she was not told, although I have heard that the Franians carry tales concerning everything that happens on the ship. Yet, she did not speak of it. It is my hope she is not aware.

I have recovered from the deep depression that hit me when I attempted to unite the council, citizens, and crew against the High One. I realize that I failed miserably in that endeavor. Prana is far cleverer than I thought her to be. She has left me humbled and deprived of the converts I formerly had.

I realigned my vision and resolved to prick her during the hour of victory when we imparted retribution on the Crezand. I was certain that the High One would break down in a womanly fashion when we sent out our death bombs. I'd believed that I would carry the day, holding her in my arms, caressing her long flow of hair, feeling the warmth of her small, shapely body. But she was a warrior. She did not bend, although I could read her distaste.

As firmly as Dar would have, she ordered mass destruction without a quiver. Then, afterward, when the bombs had sent their layer of Nox IceMist, and we watched the result spread across buildings, facilities, and the faces of the people, she remained resolute and without tears. I even tortured her with Vids, displaying the scenes of the locals' grief, but she didn't twitch an eye. She viewed them and said nothing.

What a woman. She could rule the universe. She was meant to be mine. I must have her. I must.

Shnize

Prana has done well. In the After, Dar must feel great pride in her. I tell her this, yet it does not ease her grief or her guilt. I hold her through the night and late into the morning. I praise her, but she does not heed it. I speak of the look in Terriff's eyes when she is gallant and steadfast, but she turns away, not interested in Terriff. She dwells on the tears of the Crezands. It worries me. She is too gentle. But do I want her to change? It is who she is, the Queen I love.

I embrace her body all that night, and I warm her with my arms and heart. I do not attempt to drive her into passion. She is too limp with her desolation.

I feel untrained to be Prana's Drone. Because of that, I do not have the answers for her when she needs them. I do not know how to heal a bruised soul. So, I listen and attempt to soothe her mind as best I can. My poor Queen. I wish I could place the burden on my shoulders instead of hers, but I cannot.

She needs to harden herself to such things. It is her role now as the High One. She has to learn to relish the conquering, to let the violence of battle slide across her soul like water droplets dripping onto a mat, to think like a RishMorian.

Prana

Shnize was concerned when I woke that morning and refused to eat breakfast. I did not tell him that I had thrown up my dinner in the wee hours of the morning. Already he lectured me about hardening my heart to do what was necessary. He was right, but I was not Dar, nor could I ever be.

It had been a short night. My nightmares, my nausea, and my tears had warred against all sleep, but in spite of that, I was glad to be driven forward by obligation — even, one could say, content because I was due in the War Room. After all, duty kept me from thinking about and seeing the scenes of battle, ones that kept flashing into my mind. Crezand faces looked nothing like the RishMorians' or like humans,' but Crezand tears of grief and the way their faces bunched tight in grimaces of pain — those were recognizable. The Vids had captured it all, the true horror of war.

Pushing my memories aside as much as possible, I steeled myself again. That day, I pretended that it hadn't affected me. And when the others looked at me, I wore the face that Dar had trained me to show. Thus, I displayed nothing of what I felt. I fooled all of them, all of them except Shnize.

We watched the Crezand preparations. Their retaliatory strikes were crushed with an easy nod from me. Three more of the Nox IceMist bombs showered both enemy ships and crew with frozen waste. More weeping, more grief, more anger followed. Again and again throughout the day, the Crezands attempted their defense. But they had no chance against us. RishMorian technology followed their every movement and expunged every defense.

By late afternoon, the Crezand negotiators were desperate to communicate with us. It was exactly as the War Lord had foretold. We let them brew in their rebellious frustration overnight. The next day I issued demands.

It was then that I called in the expertise of the Minister of Conclude. Misreck, a man I had never seen since I hadn't wanted to attend the conferences over Mada's marriage, and, of course, hadn't been involved in my own, entered the War Room with a great blustering of servitude.

His eyes swept over Terriff and passed by him, almost like an insult. To Jertot, he nodded stiffly. Shnize he ignored, but me, he fawned over gushingly. I accepted the obeisance and got down to business. I had a new stipulation. My Crezand husband must be a volunteer.

"This is exactly as I predicted," Terriff declared, smirking as his contemptuous sneer slid over Misreck. "You have lost your courage, High One. I do not blame you. You are not a man. However, it is obvious that I must take this onerous deed from…"

"Silence," I told him, watching his mouth clamp shut and his glare turn even more hostile. "I demand it and henceforth require this. The man who is to be my husband must come willingly, whether as sacrifice to his people or because he wants the rewards that will come when he returns to rule his planet."

Have you thought this over, my Queen? To change ships while at war is not a safe habit, Shnize mind sent.

I had never heard that expression before, but it did make sense. Yet, I had no intentions of altering my decision. There would be no drugging of the man I married, no hysterics, and no forced rape using chemical aphrodisiacs.

I could see that the others were appalled by my modification, but Minister Misreck gave me an approving look and then laughed good-humoredly. "I agree perfectly with the High One."

He bowed low and, gazing up at me, he said, "High One, you are just as astute as you are beautiful and wondrously enchanting."

I swallowed hard and sighed. "Thank you, Minister Misreck. Please rise. We have much work to do."

As if I had not spoken, the man continued. "Yes, oh wondrous High One, the willingness of a husband is to be greatly desired. Oh, listen to me. I have done a double entendre. To be desired — how appropriate," he said, laughing.

Inwardly, I groaned. Not another male bent on propositioning me. This one's eyes were drinking me in, his lips smacking, and his ears curled about his cheeks in the RishMorian prelude to lust. It was disgusting. The War Room was as close as Minister Misreck was ever going to get to me. The man was fat and old, old, old, and I hated his droopy jowls.

"Yes," I said, coldly. "How amusing."

"Ah, Sweetness of Womanhood, to have an unwilling husband could be, in fact, even dangerous, as our former High One learned. I could not bear to think of you being harmed by one of these repulsive Crezands. I shudder even at the necessity of allowing one to touch that lovely, soft, white skin of yours."

"Minister…"

"Although I am sure your guard will be in attendance, even during the wedding rites, still it would be much more preferable if your husband were eager for that great honor," he continued, cutting me off. "If would be better if his hands caressed that luscious body of yours with eager…"

"Yes, Minister Misreck. I am glad you agree with me…" I said, attempting to stop the flow.

"Although I do not understand how even such an ugly insect monster could not yearn to take you into his arms, to kiss your lips, to…"

"Minister Misreck…" I tried again, shooting a help me glance at Shnize.

"Because having observed your beauty close up like this, I am…"

"Discontinue such thoughts. You must address the High One with respect," Shnize interrupted, abruptly halting the man's flow.

Minister Misreck's eyes grew large and troubled. "I meant nothing but respect and adoration for the High One. I was just…"

"Cease these lurid descriptions," Shnize demanded, walking closer with a warning look in his eyes.

"Such a pity," Terriff said. "I was rather enjoying it."

I glared at him, but he wasn't watching me. I followed his eyes back to Misreck.

The man had finally stopped bowing. He straightened up, staring at me with the look of a lost puppy, his eyes hurt and full of rejection. "Of course," he said softly. "I had quite forgotten. Our High One is still innocent in the ways of a male. How could I have expected her to understand such things? I should never have spoken of how a man skilled in the methods of love can…"

"Enough," Shnize snarled again. This time he stepped forward, blocking me from Misreck's sight.

Terriff laughed, Jertot paled and turned several shades of yellow, as if he'd been the one offending me with inappropriate remarks. I turned away from them and walked over to the table, where I sat down. The others, including a slightly subdued Misreck, quickly joined me.

Then it was down to business. All of us had been given language boosters, and we were rapidly learning the Crezand tongue, but when negotiations began in earnest, I was very glad that I wasn't the one doing most of the talking. The clicking gutturals were far too rapid for me to follow. I relied greatly on the translating machine to verify my understanding.

The conference with the Crezand began and continued for hours, on and on. The insect heads gushed out elaborate evasions. Excuses poured forth, and then came the protestations about their innocence. It took up most of the day. I called an end to it around dinnertime and tiredly adjourned the meeting. It was just as well. Negotiations continued for the better part of a week. And then, finally, an agreement was drawn up.

That first night, again, I hardly slept. Shnize attempted to relax me in ways that were delicious, but even completion didn't still my racing heart or my fear of the next day.

I wasn't in attendance when my new husband was flown up to us. Purposefully. I stayed away, letting him become acquainted with the RishMore and with the ship. A single painting occupied his room. The painting was of me. I hoped it would help to allay his fears, or at least, to prepare him for what would come.

Shnize

I witness the Crezand husband's arrival. He is an ugly cuss, but otherwise, seems rational. He does not fight the examinations the RishMorians use to screen him inside and out. Zoff employs varied methods of exploration. None of them, I want to try out, but the volunteer placidly endures. The only question he keeps asking is "When will I get to meet my bride?"

If I have a choice in the matter, that is never. I grit my teeth. I hate this man.

Yesterday, Prana stops by to talk to Zoff. I almost laugh when she questions him about physical compatibility. Does she think we allow her to come to harm?

She blushes, turning the most interesting shade of red. Her cheeks color vibrantly. I touch one to feel the change, but her skin feels no hotter than normal. I see her do this many times. That is the first time I am able to examine the process.

She brushes me aside and asks me to wait outside. Of course, I do not. But I pretend to walk away. She is so intent on whispering secrets to the doctor, she is not aware of my presence. Her conversation enlightens.

"Zoff, does the Crezand look like a RishMore? I mean, does... does he... look like a male... the part that... you know..."

Zoff understands what she is asking, although for some reason she continues to hedge and stutter. The doctor studies her a moment then nods his head.

"It will... it will fit?"

Zoff does not laugh. He simply nods again. "The wedding rite will cause you no pain, High One... as long as you are readied, of course."

I cannot see Prana's face. I suspect that it is red once again. She is wiggling about as if she needs to urinate, although moments before she makes use of the facilities.

"Can you give me something? A tranquilizer or something?"

Zoff, of course, is horrified, as am I. Prana needs all her facilities for the joining. It is not appropriate for her to perform them in a haze.

The waves of fear that surround her grow alarmingly. She is talking herself into hysteria. I decide that her speech with Zoff is long enough.

"High One, it is time to go," I tell her.

She nods, not thinking clearly, for there is actually no duty she needs to attend. Early that morning she visits her children, and she does not meet the Crezand until the next day. Yet, she does not argue. She thanks Zoff for his time, and comes toward me, her head down, her eyes spellbound by the patterns of the wall carpet.

I return her to her chamber, strip off her clothes, and guide her into the pool. I challenge her to several races and play with her until she runs about and swims herself into exhaustion. Then, I wrap her in a soft towel and satisfy her body. By that time, the worry lines leave her young face, and she drifts off. Then, I hold her close and watch over her as she sleeps.

Chapter Nine

Plov

I have yet to meet the grand and powerful High One they keep talking about. I wonder why. She placed a picture of herself in my room. She is ugly. Perhaps she hides away, afraid to come into the light. I stare at her each day, hoping I will be able to service her. I shall not look at her face. I shall simply climb on top of her and insert myself. Nothing else is needed. Once only, I hope, and then I can go about my business, without having to look into those strange, horrible eyes.

I am not like my cousin, Mada. She was immature. She brought destruction to our planet. I told the council that they should not have chosen her. I warned them. Yet, they did not listen. Now thousands are dead, several cities have been frozen, and their people will remain untouchable for hundreds of years.

Of course, I volunteered for this position. How could I not? I must return honor to our name. I must do what Mada should have done and weave the peace blanket that the RishMore talk of. A child, half RishMore, half Crezand. Will I be able to bear the sight of it? Will it have the proper eyes or the strange, tiny nothings of the High One?

I must learn to endure the sight of such an offspring. I must learn to love it. How else can I bring it back to Crezand and allow it to grow and learn to rule our world?

The RishMore I have encountered so far have been kind. My cousin killed their leader and many, many more, yet these RishMorians have not treated me as an enemy. Many of them bow to me, awarding me the honor which is not yet mine. When will I see the High One?

They say she was the leader's wife. She must hate me. What a sacrifice it must be for her to bed me. I must view the situation in that way. I will try, anyway.

Perhaps she will find me displeasing. There is that possibility. Yet, surely she will see the beauty of my multi-hued eyes. My eyes are my best feature. Many females have told me this. I have been admired greatly. Will she feel differently?

I think sadly sometimes of the female I had once chosen as my bride. I watched her walk by my house every day. Her figure was fine, her eyes were a lovely pattern of bright greens and golds, but it was the sweetness of her smile that attracted me.

Now Cresey will never know how much I yearned to speak to her. I'll never honor my dreams of her. How sad that I did not speak to her before. I never built up the courage to tell her my thoughts. Now, she may never know I loved her.

Someday, perhaps… When I am husband of the High One, the RishMore will return me to rule. Then, I have been told, I will have the freedom to choose a companion. I hope Cresey will still be unpromised when I return. Will she wait for me? Will my heart still thump each time I view the beauty of her smile?

I scarcely dare to think of her. Yet, the days are long. Why haven't I seen my bride? What is SHE afraid of?

Shnize

"You cannot put it off any longer, my darling Prana. It must be done today," I tell her as she is brushing her already shiny hair.

She stops, frozen, her hand in midair. "Today? Not today, Shnize. Tomorrow maybe, but not…"

"You have said that for seven days now. You MUST meet him without further delay."

Prana's brush drops from her fingers. She does not notice. She turns to look at me, her face pale as the plump whiteness of a huge, puffy cloud.

"I can't, Shnize. I thought I could go through with it, but I just can't. Don't you understand?"

Her eyes are pools of reflection. They speak of the same feelings I hold inside me, but I cannot tell her that. I breathe in deeply and hold onto the air for a moment, attempting to calm myself.

"Prana, you MUST do this. Remember your sons. If you do not, Terriff will seize your position. Will that make Dar happy in the After? Do you not fear he will cut off your air from the After as punishment? You must perform your obligations."

Her eyes close, and she tilts her head upwards. A tear dangles from her right cheek. I reach out and catch it. Then I gather her into my arms.

"Prana, it MUST be done. I know you cannot bear to think of it. But I will be with you. I will watch over you. It will be quick.

Then it will be over, and he will leave. You and I can continue as we are. It is only one time with the Crezand. The marriage contract must be met, my Queen."

"I can't bear to think of it. I can't even stand to meet him. How am I going to do this?" she asks.

"Because you must. Then both of you can take care of the baby-making in the lab. But you must allow him in your bed once. That you can do. I know you have the courage."

She tries to pull away from me, but I do not permit it. There can be no more stalling.

"Besides," I tell her. "You are not being bedded today. Today you only meet him, dine with him, show that you are agreeable to the process. The RishMorians wait for you to do so. They worry, my Queen."

"I know," she cries out, but she does not attempt to pull away. She sags into me, needing my strength.

"TODAY you will welcome him to the ship, my Queen. You know the customs. You know how to do it. You must demonstrate that the two of you are together. The people need that much from you. TODAY, Prana."

Tears rain down from her eyes, wetting my chest. She clings to me, her hands like little birds nesting at my neck. I like the feel of it, the touch of her hands, her warm little body.

But there is no time to savor such things. I must work fast before she becomes argumentative again.

"Frick," I call out.

He runs into the room, as always ready to defend her. His eyes reflect disappointment when he sees that nothing is amiss. He bows and waits to hear our needs.

"Send a Franian to Plov," I order. "Inform him that the High One will see him this evening at dinner. Order him to be there."

Frick looks at Prana. "High One?" he asks, but she does not respond. She is still sobbing into my shirt. Frick stands there for several moments, hoping she will rescind the order. He does not like my dominance over the Queen. He darts worried looks from me to the High One and back again, and then finally withdraws from the room.

Prana continues to cry. I hold her like I do the boys sometimes, rocking her on my lap. Then after a time, I carry her, fully clothed, into the pool and step down into the water.

"My dress," she cries out, suddenly aware as the liquid covers her skin. "What are you doing?"

But it is okay. She is laughing. We swim and play for a while. Finally, I see that Prana is relaxing. I carry her to the bed and soothe her in the best way I know how.

Prana

Shnize was right. Each night I heard Dar calling to me, reminding me of my duty. I pushed him away and rolled over, but he woke me several times. Guilt has ridden me into exhaustion. Now Shnize set the meeting in motion. I felt ill with the thought of it, but I must go to the dining hall and meet my new husband. But oh, my poor, dead husband… Dar, I so much didn't want to do this.

Plov

I was to meet her at last. I dressed in my best. I was shown how to utilize the clothing machine, but I was determined to wear my native dress. Let her meet me as a proud Crezand. Let her see who I am from the first, I vowed to myself.

The guards escorted me. I walked slowly, trembling. I hoped no one could see. When I arrived, the High One was not there. I sat down, appearing as if I did not care, but my heart beat the pulse of anger. I lifted my goblet in preparation for the first sip of wine, and then I saw her. She was like her picture, yet not. For in her carriage, she wore the same pride as I. I saw the mirror of it in her eyes, eyes that repelled me.

I looked away. Yet, in my glance I'd seen that she was trembling. The afterimage displayed itself in the chasms of my eyes. I saw the unsteadiness of her quivering bottom lip, although her face pretended impassivity. I was suddenly drawn to the courage of the woman, drawn by her tremble. I stood up and bowed.

"Welcome," she said to me as her voice came through the translator. I wondered briefly what she sounded like without it. I stretched out my hand in a Crezand greeting. She was unsure and faltered, forgetting the Crezand greeting. I smiled at her and started to explain our customs, but she rallied and then responded appropriately, slapping the top of my hand. I laughed and slapped hers.

"We must sit down," she said, her eyes scanning the room full of staring eyes, "unless you care to go get your meal first?"

I did prefer it, but I would not tell her so. I remained at her side, sitting when she did.

"Please forgive me for not meeting you sooner," she apologized. "I am still in mourning for my husband."

I watched her face flare into various colors of red. It was quite a beautiful display. I wondered at its meaning, but she interrupted my thoughts with further words.

"I am sorry. I should not have said that."

She was the conqueror, the High One of the RishMore, yet she apologized to me. I was unsure how to react. I remained silent, pondering. She spoke again.

"What do you like to do in your spare time?" she asked, taking a sip of the wine the servant had just poured.

I picked up my goblet and copied her actions. It was safer than forging forward. Yet I saw her tiny eyes strangely begin to tear.

I cannot bear a woman's tears. I rushed forward with my response. "I walk among the Seloft trees."

Her brow wrinkled in puzzlement. It was an expression very similar to that of a confused Crezand. I understood that the translator had failed to make my words clear.

"Walk, I, in among near Seloft tall, barked things," I told her, attempting to speak the RishMorian tongue.

She laughed softly, but then she smiled. My heart beat faster. I took an extra breath and let it out slowly. When the High One smiled, it was just like Cresey's smile. I opened my mouth and returned her flash of teeth.

Unfortunately, it frightened her. I saw her recoil, but she was very brave, as I'd noticed before. She did not pull away. I reached out and took her hand. Her face paled into a lovely shade of Deftre petals. I inhaled sharply, entranced.

I had thought her ugly, and it is true that her eyes were tiny and shapeless, yet…

The people began screaming. "Hoopla, hoopla!" Their noise pulled me away from my fascination with the High One. I looked out over the crowd and nodded, acknowledging their appreciation.

Then I glanced back at my bride. My eyes scanned her shape — good-looking limbs, saucy breasts, a waist as small and dainty as Cresey's — my lips drew back, and I widened my smile. I was content. It would be far easier to fulfill my duty than I had imagined. Let the wedding begin.

Prana

He was as ugly as a giant fly, yet he seemed kind. I stared at his smile and thought about Dar, but I knew what my husband would say. *Let go of your fear, Prana. Open to the challenge.*

Plov was smiling at me, and I couldn't help grinning back. But when he took my hand, I almost bolted away.

The servant brought a plate of food for each of us. He set it down, and Plov and I both looked at the food. Then I saw from the corner of my eye that Plov was using his fingers and lifting up a piece of fruit. Slowly he carried it to my mouth. I only hesitated a moment. I understood. Plov knew our marriage customs. I opened my mouth, and he gently inserted the food.

"Hoopla, Hoopla!" the crowd clapped and cheered.

I picked up a piece from my plate and traveled it to Plov's mouth, and then I moved my hand to touch him on the arm.

"Hoopla, Hoopla!" the RishMorians roared.

We had only one more chasm to leap. He knew it, as did I. He nodded my way and leaned toward me. I nodded to him and leaned also. Then our lips met, and the token kiss was over.

"Hoopla, Hoopla!" everyone cried out while Plov gazed at me, and I at him. We had passed into the wormhole. The trip between us had begun.

"Tomorrow?" I asked, and he bowed his head.

"I am honored, High One," he said. "Marriage to you will be a pleasure."

We began to eat then, and the hall quieted down. The nobles continued to stare, but they were also dining. I glanced at Shnize. His eyes were sad, but he nodded at me. Then he smiled. *You do well, Prana. Dar in the After is proud.*

I inhaled sharply, and my eyes rose to the plaque I had ordered for the wall. *I do this for you, Dar*. I whispered. *For you and for our sons.*

Continued in ***Lost in the Sky***,
Book 3 of the RishMore

Made in the USA
Monee, IL
04 November 2026

67381187R00105